Masaaki Yoshida

Fuchsian Differential Equations

Aspects of Mathematics
Aspekte der Mathematik

Editor: Klas Diederich

Vol. E1: G. Hector/U. Hirsch, Introduction to the Geometry of Foliations, Part A

Vol. E2: M. Knebusch/M. Kolster, Wittrings

Vol. E3: G. Hector/U. Hirsch, Introduction to the Geometry of Foliations, Part B

Vol. E4: M. Laska, Elliptic Curves over Number Fields with Prescribed Reduction Type

Vol. E5: P. Stiller, Automorphic Forms and the Picard Number of an Elliptic Surface

Vol. E6: G. Faltings/G. Wüstholz et al., Rational Points
(A Publication of the Max-Planck-Institut für Mathematik, Bonn)

Vol. E7: W. Stoll, Value Distribution Theory for Meromorphic Maps

Vol. E8: W. von Wahl, The Equations of Navier-Stokes and Abstract Parabolic Equations

Vol. E9: A. Howard, P.-M. Wong (Eds.), Contributions to Several Complex Variables

Vol. E10: A. J. Tromba, Seminar on New Results in Nonlinear Partial Differential Equations
(A Publication of the Max-Planck-Institut für Mathematik, Bonn)

Vol. E11: M. Yoshida, Fuchsian Differential Equations
(A Publication of the Max-Planck-Institut für Mathematik, Bonn)

Band D1: H. Kraft, Geometrische Methoden in der Invariantentheorie

Masaaki Yoshida

Fuchsian Differential Equations

With Special Emphasis on the Gauss-Schwarz Theory

A Publication of the Max-Planck-Institut für Mathematik, Bonn
Adviser: Friedrich Hirzebruch

Springer Fachmedien Wiesbaden GmbH

Professor *Masaaki Yoshida*
Kyushu University, Fukuoka, Japan

AMS Subject Classification: 35 R 25, 35 R 30, 45 A 05, 45 L 05, 65 F 20

1987
All rights reserved
© Springer Fachmedien Wiesbaden 1987
Originally published by Friedr. Vieweg & Sohn Verlagsgesellschaft mbH, Braunschweig in 1987.

No part of this publication may be reproduced, stored in a retrieval system or transmitted in any form or by any means, electronic, mechanical, photocopying, recording or otherwise, without prior permission of the copyright holder.

Produced by W. Langelüddecke, Braunschweig

ISSN 0179-2156

ISBN 978-3-528-08971-9   ISBN 978-3-663-14115-0 (eBook)
DOI 10.1007/978-3-663-14115-0

Contents

Introduction
Notations

Part I

Chapter 1   Hypergeometric Differential Equations 1
  § 1.1   Hypergeometric Series 1
  § 1.2   Hypergeometric Equations 2
  § 1.3   Contiguity Relations 3
  § 1.4   Euler's Integral Representation 5
  § 1.5   Barnes' Integral Representation 11
  § 1.6   Confluent Hypergeometric Equations 12

Chapter 2   General Theory of Differential Equations I 14
  § 2.1   How to Write Differential Equations 14
  § 2.2   Cauchy's Fundamental Theorem 15
  § 2.3   Monodromy Representations of Differential
          Equations......................................16
  § 2.4   Regular Singularities 18
  § 2.5   The Frobenius Method 20
  § 2.6   Fuchsian Equations 24

Chapter 3   The Riemann and Riemann-Hilbert Problems 27
  § 3.1   Statement of the Problems 27
  § 3.2   An Observation 28
  § 3.3   The Solution of the Riemann Problem 29
  § 3.4   Apparent Singularities 29
  § 3.5   A Solution of the Riemann-Hilbert Problem 30
  § 3.6   Isomonodromic Deformations 35

Chapter 4   Schwarzian Derivatives I 38
  § 4.1   Definitions and Properties 38
  § 4.2   Relations With Differential Equations 39

§ 4.3   A Canonical Form of Differential Equations 42
§ 4.4   Local Behaviour of Schwarzian Derivative 44

Chapter 5   The Gauss-Schwarz Theory for Hypergeometric
            Differential Equations 46
§ 5.1   Orbifolds and Their Uniformizations 46
§ 5.2   Uniformizing Differential Equations 48
§ 5.3   The Gauss-Schwarz Theory 51

Part II

Chapter 6   Hypergeometric Differential Equations in
            Several Variables 58
§ 6.1   Hypergeometric Series 58
§ 6.2   Hypergeometric Differential Equations 60
§ 6.3   Contiguity Relations 63
§ 6.4   Euler's Integral Representations 65
§ 6.5   The Erdelyi-Takano Integral Representations 78
§ 6.6   The Barnes Integral Representations 78
§ 6.7   A Relation Between the Equation F_D and
        the Garnier System G_N 79
§ 6.8   Confluent Hypergeometric Equations 81

Chapter 7   The General Theory of Differential Equations ...82
§ 7.1   Singularities of Differential Equations 82
§ 7.2   Holonomic Systems 85
§ 7.3   The Equation F_1 89
§ 7.4   Equation of Rank $n+1$ 90
§ 7.5   Differential Equations on Manifolds 92
§ 7.6   Regular Singularities 93

Chapter 8   Schwarzian Derivatives II 95
§ 8.1   Definitions and Properties 95
§ 8.2   Relations With Differential Equations 99
§ 8.3   A Canonical Form of Differential Equations 100
§ 8.4   Projective Equations and Projective
        Connections 102
§ 8.5   Local Properties of Schwarzian Derivatives 104

Chapter 9   The Riemann and Riemann-Hilbert Problems II ..107
  § 9.1 The Riemann Problem in Several Variables 107
  § 9.2 Accessory Parameters 110

Chapter 10 The Gauss-Schwarz Theory in Two Variables 115
  § 10.1 Uniformizability of Orbifolds 116
  § 10.2 Orbifolds Whose Universal Uniformizations
         Are Symmetric Spaces 118
  § 10.3 Line Arrangements in P_2 and Orbifolds
         Uniformized by B_2 124
  § 10.4 Uniformizing Differential Equations 135
  § 10.5 The Gauss-Schwarz Theory for Appell's
         Equation F_1 137
  § 10.6 The Monodromy Representation of Appell's
         Equation F_1 145

Chapter 11 Reflection Groups 154
  § 11.1 Unitary Reflection Groups 155
  § 11.2 Unitary Reflection Groups of Dimension 2 159
  § 11.3 Parabolic Reflection Groups of Dimension 2 ...170
  § 11.4 Line Arrangements Defined by Unitary
         Reflection Groups of Dimensions 3 and 4 179

Chapter 12 Toward Finding New Differential Equations 181
  § 12.1 Tensor Forms Associated to the Differential
         Equations 181
  § 12.2 Integrability Conditions 185
  § 12.3 The Restricted Riemann Problem 188
  § 12.4 G-Invariant Differential Equations I 191
  § 12.5 G-Invariant Differential Equations II 198

References 204

Sources of Figures 215

Relations of Chapters

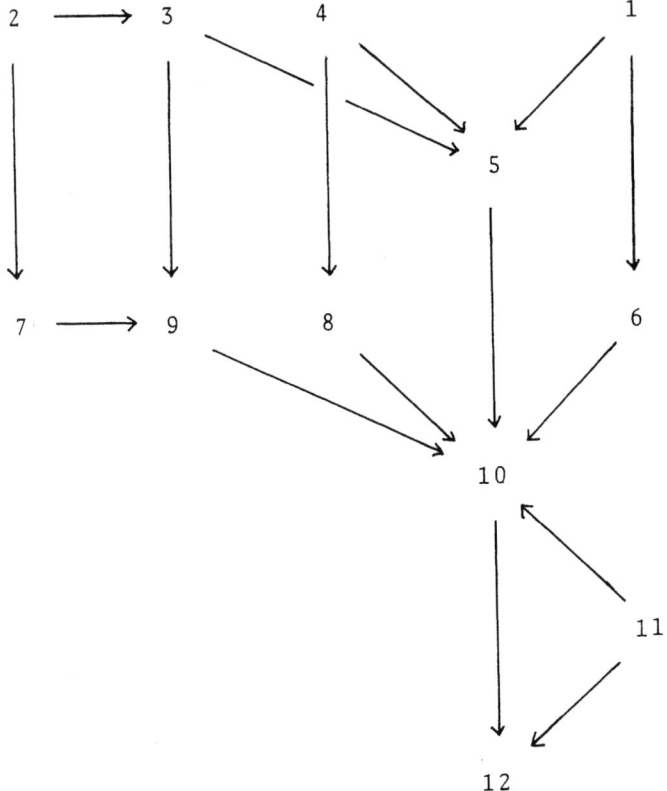

"p → q" means "consult Chapter p when reading Chapter q"

Introduction

This book stems from lectures given at the Max-Planck-Institut für Mathematik (Bonn) during the winter semester 1985/1986 under the title "Fuchsian Differential Equations".

The aim of this series of lectures was to study linear ordinary differential equations and systems of linear partial differential equations with finite dimensional solution spaces in the complex analytic category; referred to in the course of this book as differential equations. This field has a long history and has attracted many famous mathematicians such as Euler, Gauss, Schwarz and Kummer, for example. Cauchy, L. and R. Fuchs, Painleve, Riemann, Poincare and Birkhoff were also interested in these equations. In many text books, as for instance in [Bie], in [Huk] and in [Inu], one can find the classical results in this direction, while the most recent ones are to be looked for in the book edited by Gérard and Okamoto [G-O].

The present work does not intend to be a new textbook discussing all aspects of the subject. On the contrary it concentrates exclusively on one particular but essential point, namely, the Gauss-Schwarz theory which deals with the hypergeometric differential equation. This theory is of special interest because its development was the occasion for many fruitful interactions between differential equations and other branches of mathematics, for example, algebraic geometry, group theory and differential geometry. For this reason, we hope that our book will give an opportunity to mathematicians working in other fields to get interested in differential equation theory — which can undoubtedly be most profitable for their own work. The interrelations mentioned above originated partly in the age of Euler and Gauss, but have been rather overlooked in the recent past.

Here the Gauss-Schwarz theory is presented as part of the theory of orbifolds due to Satake and Thurston. To each orbifold uniformized by a symmetric space (i.e. an orbifold whose universal cover, in Thurston's sense, is a Hermitian symmetric space), it should be possible to associate what is called here a uniformizing (differential) equation. In the two main cases, it is indeed possible to do so as explained in the sequel, but in the remaining situations the possibility is still a conjecture. In the present context, the hypergeometric equation turns out to be the uniformizing equation attached to the orbifold on the complex projective line with three singular points. In a higher dimensional generalization of the Gauss-Schwarz theory, Appell's hypergeometric equation F_1 in two variables appears as the uniformizing equation of the orbifolds on the complex projective plane with singularities along the complete quadrilateral. Other examples are also constructed here; they uniformize orbifolds on the complex projective plane with singularities along some line arrangements with high symmetry properties.

At first glance one would assume it is easy to find non-trivial differential equations with given singularities. It is true in the single variable case, but in several variables it is quite a hard problem because the coefficients of such an equation must satisfy a system of non-linear differential equations, called the integrability condition. Till now, no efficient theory has been developed to tackle this problem. Since our construction of uniformizing equations is restricted, our approach should be considered as a partial answer to this problem.

By studying this theory one encounters numerous interesting topics : some of them are so attractive that we could not refrain from developing them here. This is the case, for instance, in § 1.3, 1.5, 1.6, 3.6, 6.3, 6.5, 6.6, 6.7 and in § 6.8. We have tried our best to ensure that mathematicians not familiar with differential equations and even undergraduate students may enjoy reading the present text. For this reason no knowledge of differential equations is necessary to understand the main theory. On the way, however, we meet several other branches of mathematics where some theorems on number theory, differential geometry, algebraic geometry, topology and group theory are quoted without proofs but always with references (Chapters 10 and 11). Moreover in § 3.5 where a short digression is made, it is assumed that the reader is familiar with the foundations of algebraic geometry.

The material of this book is organized as follows.
Part I treats ordinary differential equations. In Chapter 1 the classical hypergeometric differential equation is studied in detail. It is not so much because our presentation is intended to be elementary but essentially since this differential equation provides the whole theory with a leading example. Indeed one will encounter in the course of the book the deep meaning of each elementary classical fact. Although no general theory is going to be developed, some basic ingredients of ordinary differential equations are needed: these are recalled in Chapter 2. Chapter 3 studies the Riemann problem — the problem of finding differential equations with given local behaviour — and the Riemann-Hilbert problem — the problem of finding those equations of which monodromy representations are given as representations of the fundamental group of the complement of the given divisors. The first problem is almost trivial in the single variable case but very difficult in the several variable case. The second problem is, in some sense, reasonably easy in both situations. The Schwarzian derivative defined in Chapter 4 plays an important role in the succeeding chapter. Chapter 5 introduces the concept of orbifolds and studies the classical Gauss-Schwarz theory from the viewpoint of the uniformizations of orbifolds.

Part II treats differential equations in several variables. Chapter 6 introduces Appell's hypergeometric differential equations. Numerous and detailed formulae are given since as far as we know there are only a few books on hypergeometric equations in several variables: for example, the classical book [A-K], the collection of formulae [Erd] and the elementary lecture note [Kim 1]. In Chapter 7 some basic facts on partial differential equations are recalled. Chapter 8 generalizes the Schwarzian derivative to several variables, this generalization plays a key role in Chapters 10 and 12. In Chapter 9 the Riemann and the Riemann-Hilbert problems in several variables are discussed; the situation looks totally different from that in the single variable case. Chapter 10 introduces the notion of uniformizing differential equations for 2-dimensional orbifolds. It is not easy to obtain 2-dimensional orbifolds uniformized by symmetric spaces (the case we are interested in). In order to construct such orbifolds recent and very deep results of R. Kobayashi are used in a crucial way. We do not give proofs since these would require an entire volume, but if the reader is willing to believe the theorems the understanding is in no way impaired. The Gauss-Schwarz theory for hypergeometric differential equation F_1 in two variables is then studied in detail.

Chapter 11 studies various reflection groups and shows that they are deeply connected with orbifolds. It is intended to provide the reader with a good understanding of Chapter 10 and works as a transition to Chapter 12. In Chapter 12 at which all the book is aiming, new Fuchsian differential equations in two variables which have remarkable properties are constructed. These equations are highly symmetric and for special values of the parameters they turn out to be the uniformizing equations of certain orbifolds on the projective plane.

At the end the author would like to thank Professors F. Hirzebruch and G. Wüstholz who offered him the opportunity to give the series of lectures that gave rise to this book. He is also grateful to his friends : R. Endell, A. Flegmann, R. Gérard, T. Höfer, B. Morin, K. Okamoto and J. Werner.

Kyushu University, January, 1987

Masaaki YOSHIDA

Notations

$\mathbb{Z}$	ring of integers
$\mathbb{Q}$	rational number field
$\mathbb{R}$	real number field
$\mathbb{C}$	complex number field or the complex line
$\mathbb{C}^*$	multiplicative group $\mathbb{C} - \{0\}$
$R[x_1,\ldots,x_n]$	polynomial ring in n indeterminates over a ring R
$K(x_1,\ldots,x_n)$	rational function field in n indeterminates over a field K
$O(U)$	ring of holomorphic functions on an open subset U in $\mathbb{C}^n$.
$D(U)$	ring of linear differential operators with coefficients in $O(U)$
Re z	real part of a complex number z
Im z	imaginary part of a complex number z
$\mathbb{C}^n$	complex affine space of dimension n
P_n	complex projective space of dimension n
H	upper half plane $\{\, z \in \mathbb{C} \mid \operatorname{Im} z > 0 \,\}$
B_n	complex n dimensional ball $\{\,(z_1,\ldots,z_n) \in \mathbb{C}^n \mid \lvert z_1 \rvert^2 + \ldots + \lvert z_n \rvert^2 < 1 \,\}$
dim V	dimension of a vector space V over $\mathbb{C}$
det A	determinant of a matrix A
tr A	trace of a matrix A
${}^t A$	transpose of a matrix A
End(V)	ring of endomorphisms of a linear space V
GL(n,R)	general linear group of dimension n over a ring R
PGL(n,R)	projectivization of GL(n,R)

$SL(n,R)$	$= \{ X \in GL(n,R) \mid \det X = 1 \}$
$PSL(n,R)$	projectivization of $SL(n,R)$
$U(n)$	unitary group of dimension n
$G \ltimes H$	simi-direct product of groups G and H (G acts on H)
$E(n)$	group of complex euclidean rigid motions $U(n) \ltimes \mathbb{C}^n$
$U(n,1)$	unitary group of signature $(n,1)$
$PU(n,1)$	projectivization of $U(n,1)$
$<p,q,r>$	polyhedral or triangle group
$Aut(X)$	group of automorphisms of X
S_n	symmetric group on n letters
$H^p(X,S)$	p-th cohomology group of X with coefficients in a sheaf S
$O(V)$	sheaf of germs of holomorphic sections of a vector bundle V
Ω^p	sheaf of germs of holomorphic p-forms
$\Pi_1(X,a)$	fundamental group of X with base point a
δ_i^j	Kronecker's symbol
$\{z;x\}$	Schwarzian derivative of z with respect to x
$S_{ij}^k(z;x)$	Schwarzian derivatives of z with respect to x of dimension n (≥ 2)
$\Gamma(x)$	Euler's gamma function
HGS	hypergeometric series
HGDE	hypergeometric differential equation
HGF	hypergeometric function
F_j	Appell's HGS , HGDE or HGF ($j = 1,2,3,4$)
F_J	Lauricella's HGS, HGDE or HGF ($J = D,A,B,C$)
$\odot$	symmetric tensor
PTMD	Picard-Terada-Mostow-Deligne
# Y	cardinality of a set Y
$c_j(Y)$	j-th Chern class of a manifold Y
$c_j(X,b)$	j-th Chern class of an orbifold (X,S,b)
$<a,b,...>$	group generated by $a,b,...$

Part I

Chapter 1 Hypergeometric Differential Equations

In this chapter the topic of this book is presented. One considers first the hypergeometric series which gives rise to the celebrated hypergeometric equation. The integral representations of this equation are then studied. In addition the contiguity relations and the phenomenon of confluence are mentioned.

§ 1.1 Hypergeometric Series

For any complex number a and any integer n we denote by (a,n) the product

$$(a,n) = a(a+1)\cdots(a+n-1).$$

In particular, we have $(1,n) = n!$. Furthermore we have

$$(a,n) = \frac{\Gamma(a+n)}{\Gamma(a)}$$

where Γ denotes the Gamma function.
Following Euler(1769), for any complex numbers a, b and c ($c \neq 0,-1,-2\cdots$) let us consider the power series

$$F(a,b,c;x) = \sum_{n=0}^{\infty} \frac{(a,n)(b,n)}{(c,n)(1,n)} x^n$$

This power series is called the hypergeometric series (in short, the HGS). Put

$$A_n = \frac{(a,n)(b,n)}{(c,n)(1,n)}$$

then we have

$$\frac{A_{n+1}}{A_n} = \frac{(a+n)(b+n)}{(c+n)(1+n)}$$

If a or b is a non positive integer, then $F(a,b,c;x)$ is a polynomial in x; in the other cases, since

$$\frac{A_{n+1}}{A_n} \to 1 \quad \text{as} \quad n \to \infty,$$

the series converges exactly in the open unit disc and has at least one singularity on the unit circle.

§ 1.2 Hypergeometric Equations

We now construct a linear differential equation for which $F(a,b,c;x)$ is a solution. Put

$$D = x\frac{d}{dx}$$

and notice that, if f is a polynomial, we have

$$f(D) x^n = f(n) x^n.$$

Using this formula and the above expression of A_{n+1}/A_n, we see that

$$\{(a+D)(b+D) - (c+D)(1+D)x^{-1}\}(\sum_{n=0}^{\infty} A_n x^n)$$

$$= \sum \{(a+n)(b+n)A_n x^n - (c+n-1)(1+n-1)A_n x^{n-1}\}$$

$$= \sum (a+n)(b+n)A_n x^n - \sum (c+n)(1+n)A_{n+1}x^n$$

$$= 0$$

which proves the following result:

<u>Proposition</u>: As a function of x, the HGS $F(a,b,c;x)$ satisfies the following second order linear differential equation in the unknown u

$$\{(a+D)(b+D) - (c+D)(1+D)x^{-1}\}u = 0.$$

This equation is equivalent to the so called hypergeometric differential equation (the <u>HGDE</u> for brevity):

$$x(1-x)\frac{d^2u}{dx^2} + \{c - (a+b+1)x\}\frac{du}{dx} - abu = 0$$

It has singularities at $x = 0, 1$ and ∞. A solution of the HGDE is called a hypergeometric function (in short, an <u>HGF</u>).

§ 1.3 Contiguity Relations

The following functions

$$F(a\pm 1,b,c,x), \quad F(a,b\pm 1,c;x) \quad \text{and} \quad F(a,b,c\pm 1;x)$$

are said to be contiguous to $F(a,b,c;x)$. They are linear combinations of $F(a,b,c;x)$ and of its first derivative with respect to x:

<u>Proposition</u>: (contiguity relations) We have

$$F(a+1,b,c;x) = \frac{1}{a}(x\frac{d}{dx} + a)F$$

$$F(a-1,b,c;x) = \frac{1}{c-a}\{x(1-x)\frac{d}{dx} - bx + c - a\}F$$

$$F(a,b+1,c;x) = \frac{1}{b}(x\frac{d}{dx} + b)F$$

$$F(a,b-1,c;x) = \frac{1}{c-b}\{x(1-x)\frac{d}{dx} - ax + c - b\}F$$

$$F(a,b,c+1;x) = \frac{c}{(c-a)(c-b)}\{(1-x)\frac{d}{dx} + c - a - b\}F$$

$$F(a,b,c-1;x) = \frac{1}{c-1}(x\frac{d}{dx} + c - 1)F$$

where F stands for $F(a,b,c;x)$.

Proof: Since the HGS is symmetric with respect to a and b, actually the contiguity relations for b are redundant. Using the notations in §1.1, we have

$$F(a+1,b,c;x) = \sum \frac{(a+1,n)(b,n)}{(c,n)(1,n)} x^n$$

$$= \sum A_n \frac{a+n}{a} x^n$$

$$= \frac{1}{a}(a \sum A_n x^n + \sum A_n n x^n)$$

$$= \frac{1}{a}(D + a)F,$$

and

$$F(a,b,c-1;x) = \sum A_n \frac{c-1+n}{c-1} x^n$$

$$= \frac{1}{c-1}(D + c - 1)F.$$

In order to obtain the remaining contiguity relations, we use the following equalities.

$$(\frac{d}{dx} - x\frac{d}{dx} - b)F = \sum \{A_{n+1}(n+1) - A_n n - bA_n\} x^n$$

$$= \sum A_n \{\frac{(a+n)(b+n)}{c+n} - (n+b)\} x^n$$

$$= (a-c)\sum A_n \frac{b+n}{c+n} x^n$$

Indeed we have

$$F(a-1,b,c;x) = \sum A_n \frac{a-1}{a+n-1} x^n$$

$$= F - \sum_n A_n \frac{n}{a+n-1} x^n$$

$$= F - \sum_n A_{n-1} \frac{b+n-1}{c+n-1} x^n$$

$$= F - x \sum_n A_n \frac{b+n}{c+n} x^n$$

$$= F - \frac{x}{a-c} (\frac{d}{dx} - x \frac{d}{dx} - b) F$$

and

$$F(a,b,c+1;x) = \sum_n A_n \frac{c}{c+n} x^n$$

$$= \sum_n A_n \frac{c}{b-c} (\frac{b+n}{c+n} - 1) x^n$$

$$= \frac{c}{b-c} \{\frac{1}{a-c} (\frac{d}{dx} - x \frac{d}{dx} - b) - 1\} F.$$

This ends the proof of the proposition.

<u>Remark</u>: The contiguity relations imply in particular that any three functions of the form $F(a',b',c';x)$, with the conditions $a'-a \in \mathbb{Z}$, $b'-b \in \mathbb{Z}$ and $c'-c \in \mathbb{Z}$, are linearly related over the field $\mathbb{Q}(a,b,c,x)$ of rational functions in a,b,c and x.

§ 1.4 Euler's Integral Representation

The use of Euler's integral that we are now defining gives all the solutions of the HGDE. For any complex number λ, consider the Euler kernel

$$K(\lambda,t,x) = (1 - tx)^{-\lambda}.$$

The Euler transformation $u(x)$ of a function $w(t)$ is given by the following formal integral

(1.1) $u(x) = \int w(t) K(\lambda,t,x) dt,$

where "formal" means that no path of integration has been yet specified and that no convergence has been discussed. We want to find a reasonably simple function $w_0(t)$ such that its Euler transformation is a solution of the HGDE. In order to obtain it, we need to find a linear differential operator of first order

$$Q = Q(t,\theta) \quad \text{where} \quad \theta = t\frac{d}{dt}$$

and a function $L = L(x,t)$ such that

$$PK(\lambda,t,x) = QL,$$

where $P = P(x,D) = (a+D)(b+D) - (c+D)(1+D)x^{-1}.$

Indeed with such a Q and such an L we will have, modulo Note 2 below:

$$(1.2) \quad Pu = \int w(t)PK(\lambda,t,x)dt$$

$$= \int w(t)QL\,dt$$

$$= \int Q^*w(t)L\,dt + R$$

where Q^* is the formal adjoint operator of Q and R the remainder term defined in Note 1.

By definition the formal adjoint operator Q^* of any linear differential operator

$$Q = \sum a_j(t)\frac{d^j}{dt^j}$$

is

$$Q^* = \sum (-1)^j \left(\frac{d^j}{dt^j}\right) a_j(t).$$

Note: 1) By using integration by parts several times, one sees that for any smooth functions f and g the following is true

$$\int_a^b fQg\, dt = \int_a^b (Q^*f)g\, dt + R$$

where R is a remainder term depending only on the values at a and b of f, of g, of the coefficients of Q and of some of their derivatives.

If the path can be chosen so that $R = 0$, the desired function w_0 will be any of the solutions of the equation

$$Q^*w = 0.$$

<u>Note</u>: 2) The reader must aware that the first equality sign in the previous computation (1.2) is formal since one needs uniform convergence in x along the path of integration in order that (1.1) be defined and that differentiation in the x direction under the integral sign be permitted.

<u>Construction of Q and L</u> : Notice first that for all $\lambda \in \mathbb{C}$ we have

$$DK(\lambda,t,x) = \theta K(\lambda,t,x)$$

$$(b+D)K(\lambda,t,x) = \lambda K(\lambda+1,t,x) + (b-\lambda)K(\lambda,t,x)$$

$$(1+D)x^{-1}K(\lambda,t,x) = \lambda t K(\lambda+1,t,x).$$

By setting $\lambda = b$ we get

$$PK(b,t,x) = (a+D)bK(b+1,t,x) - (c+D)btK(b+1,t,x)$$

$$= b\{(a + \theta) - t(c + \theta)\}K(b+1,t,x).$$

This computation leads us to put $Q = a + \theta - t(c + \theta)$ and $L = K(b+1,t,x)$. Now we have

$$Q = a - ct + t\frac{d}{dt} - t^2\frac{d}{dt}$$

and $$Q^* = a - ct - \frac{d}{dt}t + \frac{d}{dt}t^2$$

$$= a - ct - t\frac{d}{dt} - 1 + 2t + t^2\frac{d}{dt}$$

$$= t(t-1)\{\frac{d}{dt} - \frac{(a-1)}{t} - \frac{(c-a-1)}{(t-1)}\}.$$

Up to multiplicative constant, the equation $Q^*w = 0$ has a unique solution

$$w_0 = t^{a-1}(1-t)^{c-a-1}.$$

Therefore

$$u = \int t^{a-1}(1-t)^{c-a-1}(1-tx)^{-b} dt$$

is a formal solution of the HGDE.

Let us now specify the paths along which we want to integrate. We are not going to characterize here all the paths that lead to solutions of the HGDE, we only specify two particularly useful and well known ones.

<u>The segment</u> $(0,1)$: If

(1.3) $\text{Re } a > 0$ and $\text{Re}(c - a) > 0$

then the integral

$$(1.4) \quad \int_0^1 t^{a-1}(1-t)^{c-a-1}(1-tx)^{-b} dt$$

converges uniformly in x so that it defines a holomorphic function in x. Moreover (1.4) is a solution of the HGDE as we show now. If $\text{Re } a$ and $\text{Re}(c-a)$ are sufficiently big, one can differentiate under the integral sign in the x direction, the remainder term R in the formal computation (see Note 1 above) turns out to be zero and so integral (1.4) is a solution of the HGDE. Since it is meromorphic in a, b and c, using analytic continuation, we see that provided (1.4) remains convergent it is still a solution when a, b and c vary. The fact that

$$F(a,b,c;x) = \frac{\Gamma(c)}{\Gamma(a)\Gamma(c-a)} \int_0^1 t^{a-1}(1-t)^{c-a-1}(1-tx)^{-b} dt.$$

$$(|x| < 1, \text{ Re } a > 0, \text{ Re}(c-a) > 0, c \neq 0, -1, -2, \cdots)$$

can be directly checked by using the following formulae

$$(1 - tx)^{-b} = \sum \frac{(b,n)}{(1,n)} t^n x^n \qquad (|x| < 1)$$

and
$$\int_0^1 t^{p-1}(1-t)^{q-1} dt = \frac{\Gamma(p)\Gamma(q)}{\Gamma(p+q)} \qquad (\text{Re } p, q > 0)$$

(the latter being the formula defining the beta function)

Of course one is willing to get rid of the restriction (1.3). This can be achieved by considering the following loop (i.e. a path that ends where it starts).

<u>Double contour loop around 0 and 1</u>: Let γ be a closed curve starting from a point T in the open interval (0,1), turning in the positive direction around 0 and 1 and next in the negative direction around 0 and 1 and ending at T as shown here (double contour loop around 0 and 1).

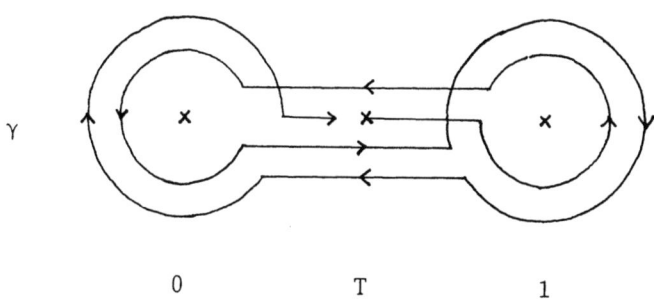

If f(t) is a function holomorphic in some neighbourhood U of the closed interval [0,1] in $\mathbb{C}$ and if γ is assumed to be inside U, then for any complex numbers p and q, the value of $t^p(1-t)^q f(t)$ is the same at the starting point and at the end point of γ as can be checked by using anaytic continuation. Thus the integral

$$\int_\gamma t^p(1-t)^q f(t)dt$$

is well defined in the sense that it is independent of the origin chosen on the loop and even of the shape of the loop provided you deform it smoothly in U - {0,1} and provided of course it remains a loop during the deformation. Since γ is compact the integral

$$u(x) = \int_\gamma t^{a-1}(1-t)^{c-a-1}(1-tx)^{-b} dt$$

converges uniformly in x so that the formal computation leading to introduce Q^* has a meaning. The remainder term R of Note 1 above vanishes since the integrand is single valued along γ, so that the integral is a solution of the HGDE. The double contour loop around any two of the four points 0, 1, 1/x and ∞ also work. Check that if a, b and c are not integers any two solutions obtained by two different double contour loops are linearly independent so that all the solutions of the HGDE can be reached. We show a picture of a double contour loop in the case when two points are 1/x and ∞.

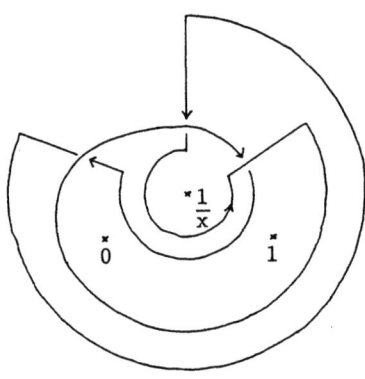

We now study the relation between these two kinds of paths. Let f be a holomorphic function f near [0,1] and γ the double contour loop around 0 and 1. When Re p > -1 and Re q > -1, by considering the variation of the argument, we see that

$$\int_\gamma t^p(1-t)^q f(t)dt = \{1 - e^{2\pi i q} + e^{2\pi i(p+q)} - e^{2\pi i p}\}$$

$$\times \int_0^1 t^p(1-t)^q f(t)dt.$$

If p and q are not integers, this gives us a method to change the integral on the segment $[0,1]$ into the integral on the double contour loop around 0 and 1.

§ 1.5 Barnes' integral representation

Consider the Mellin kernel x^s and the formal Mellin transformation $u(x)$ of a function $z(s)$:

$$u(x) = \int z(s)x^s ds.$$

The Mellin transformation u of z has the following properties:

$$(a + D)u = \int z(s)(a+s)x^s ds$$

and
$$x^{-1}u = \int z(s+1)x^s ds.$$

Let P be the linear differential operator introduced in §1.4 then we have the following computation

$$Pu = \int \{z(s)(a+s)(b+s) - z(s+1)(c+s)(1+s)\}x^s ds.$$

Therefore the transformation u of

$$z(s) = \frac{\Gamma(a+s)\Gamma(b+s)}{\Gamma(c+s)\Gamma(1+s)}\{\text{ a periodic function of period 1 }\}$$

is a solution of the HGDE provided that the integral converges and that differentiation in the x direction under the integral sign is permitted. Of course this uniform convergence condition will hold only for appropriate choices of the function z and of the path of the integration γ. For example when

$$z(s) = \frac{\Gamma(a+s)\Gamma(b+s)}{\Gamma(c+s)\Gamma(1+s)} \frac{-\pi}{\sin \pi s} \quad \text{and} \quad \gamma = [-i\infty, i\infty]$$

the residue calculus leads to

$$F(a,b,c;x) = \frac{1}{2\pi i} \frac{\Gamma(c)}{\Gamma(a)\Gamma(b)} \int_{-i\infty}^{+i\infty} \frac{\Gamma(a+s)\Gamma(b+s)}{\Gamma(c+s)\Gamma(1+s)} \frac{-\pi}{\sin \pi s} (-x)^s ds$$

$$= \frac{1}{2\pi i} \frac{\Gamma(c)}{\Gamma(a)\Gamma(b)} \int_{-i\infty}^{+i\infty} \frac{\Gamma(a+s)\Gamma(b+s)}{\Gamma(c+s)} \Gamma(-s)(-x)^s ds.$$

This is the Barnes integral representation (1908) of the HGS.

§ 1.6 Confluent Hypergeometric Equation

Last let us consider the confluent hypergeometric series

$$F(a,c;x) = \sum_{n=0}^{\infty} \frac{(a,n)}{(c,n)(1,n)} x^n.$$

It is obtained from $F(a,b,c;x)$ by using the transformation $x \to x/b$ and by letting $b \to \infty$. This series converges in the whole plane and satisfies the differential equation

$$\{(a+D) - (c+D)(1+D)x^{-1}\} u = 0.$$

This equation is equivalent to Kummer's equation

$$xu'' + (c-x)u' - au = 0.$$

The above process, obtaining the confluent HGS, merges a singular point $x = 0$ into another singular point $x = \infty$ of the HGDE. The resulting equation (Kummer's equation) has singularities only at $x = 0$ and ∞. This is the reason why this limit process is called "confluence".

An integral representation of the confluent HGS is given, for example, by

$$F(a,c;x) = \frac{\Gamma(c)}{\Gamma(a)\Gamma(c-a)} \int_0^1 t^{a-1}(1-t)^{c-a-1} e^{tx} dx,$$

since the Euler kernel $(1-tx/b)^{-b}$ converges to the Fourier-Laplace kernel e^{tx} when b tends to infinity (confluence principle).

Chapter 2   General Theory of Differential Equations I

Before we continue to study the HGDE, we collect in this chapter some basic facts about linear ordinary differential equations defined on complex domains. The knowledge of these is sufficient but also necessary to understand this book. Any reader familiar with the theory can skip this chapter.

§ 2.1 How to Write Differential Equations

We want to study interesting special classes of linear ordinary differential equations of the form

(H) $\qquad \sum_{j=0}^{r} a_j(x) \dfrac{d^j u}{dx^j} = 0$

These are equivalent to systems of first order differential equations of the form

(M) $\qquad \dfrac{d}{dx}\, {}^t(u_1, \ldots, u_r) = A(x)\, {}^t(u_1, \ldots, u_r)$

and vice versa. To develop a general theory, it is more convenient to consider (M). But when studying particular cases, contrary to what many people believe, it is awkward first to obtain results on (M) and next to translate them to (H). Indeed since the correspondence is by no means one-to-one such a procedure is always very delicate and difficult and in some cases impossible (an example is given in §2.4). For this reason we study directely equations of the form (H). Moreover we restrict ourselves to second order equations defined on the whole complex plane, because one can easily imagine what should happen in the general case — equations of higher order defined on Riemann surfaces.

§ 2.2 Cauchy's Fundamental Theorem

Consider the linear ordinary differential equation of second order:

(1) $\quad \dfrac{d^2 u}{dx^2} + p(x)\dfrac{du}{dx} + q(x)u = 0$

where p and q are rational functions of x. A point $x = x_0$ is called a singular point of (1) if p or q has a pole at x_0. The point at infinity is called a singular point of (1) if after changing variable x into $t = 1/x$ the transformed equation

(2) $\quad \dfrac{d^2 u}{dt^2} + \{ \dfrac{2}{t} - \dfrac{1}{t^2}p(\dfrac{1}{t}) \}\dfrac{du}{dx} + \dfrac{1}{t^4}q(\dfrac{1}{t})u = 0.$

has a singularity at $t = 0$.

<u>Theorem</u> (Cauchy): If $x = x_0$ is a non singular point of the equation (1), then there exist two linearly independent holomorphic solutions of (1) around x_0.

The proof of this theorem can be found in any elementary book about general calculus. Therefore we omit it here.

<u>Remark</u>: The converse of this theorem does not hold, i.e. Existence of two linearly independent holomorphic solutions of (1) around x_0 does not necessary imply that x_0 is a non-singular point of (1). (See § 2.5 and § 3.4.)

We define the <u>Wronski determinant</u> (<u>Wronskian</u>) W of two linearly independent solutions u_1 and u_2 of (1) by

$$W = W(u_1, u_2) = \det \begin{pmatrix} u_1 & u_2 \\ u_1' & u_2' \end{pmatrix}$$

where u' stands for the first derivative du/dx. The Wronskian W satisfies the following differential equation

$$\frac{dW}{dx} = -p(x)W$$

so we get

$$W(x) = c \exp \int^x -p(t)\,dt$$

where c is a constant. As u_1 and u_2 are linearly independent, W is not identically zero and therefore $c \neq 0$. As a consequence we have that if x_0 is a non-singular point of (1), then W(x) is a holomorphic function at x_0 which does not vanish at x_0.

§ 2.3 Monodromy Representation of Differential Equations

By

$$S = \{x_1, \ldots, x_m, x_{m+1} = \infty\}$$

we denote the set of singular points of (1). Let now x_0 be a point of $X = \mathbf{P}_1 - S$ and let u_1 and u_2 be linearly independent solutions of (1) around x_0. Furthermore let C be a curve in X starting from and ending at x_0. If we continue u_1 and u_2 analytically along C, the functions remain linearly independent solutions (because of the Wronskian argument above). So there is a two-by-two nonsingular matrix M(C) such that

$$C_*{}^t(u_1, u_2) = M(C)^t(u_1, u_2)$$

where C_* means the analytic continuation along C.

The matrix $M(C)$ is called <u>the circuit matrix of (u_1,u_2) along</u> C. If C_1 and C_2 are curves starting from and ending at x_0 then the composition $C_1 \cdot C_2$ of the curves is defined to be the curve following firstly C_1 and then C_2. We have

$$(C_1 \cdot C_2)_*{}^t(u_1,u_2) = (C_2)_*(C_1)_*{}^t(u_1,u_2)$$

$$= (C_2)_* M(C_1){}^t(u_1,u_2)$$

$$= M(C_1)M(C_2){}^t(u_1,u_2)$$

and so $M(C_1 \cdot C_2) = M(C_1)M(C_2)$.

If C_1 can be continuously deformed in X (fixing x_0) into C_2, then we have $M(C_1) = M(C_2)$. Therefore denoting by $\Pi_1(X,x_0)$ <u>the fundamental group of</u> X <u>with base point</u> x_0 (i.e. the group of homotopy equivalence classes of curves starting from and ending at x_0), the correspondence $C \mapsto M(C)$ induces the homomorphism

$$\Pi_1(X,x_0) \to GL(2,\mathbb{C})$$

which is called <u>the monodromy representation</u> of the differential equation (1).

This map is depending on the choice of u_1 and u_2 as well as x_0. If we change u_1 and u_2, the new representation is conjugate to the old one. It is also the case when we change x_0. So the equation (1) determines the conjugacy class of a representation $\Pi_1(X) \to GL(2,\mathbb{C})$.

<u>The monodromy group</u> of (1) is now defined as the image of the monodromy representation in $GL(2,\mathbb{C})$. <u>The projective monodromy group</u> of (1) is defined to be the image of the monodromy group under the natural map $GL(2,\mathbb{C}) \to PGL(2,\mathbb{C})$. The conjugacy class of the monodromy group in $GL(2,\mathbb{C})$ and the conjugacy class of the projective monodromy group are determined uniquely by the equation (1).

Remark: In Chapter 11 we give the explicit monodromy representation of the HGDE.

§ 2.4 Regular Singularities

Let $x = 0$ be a singular point of (1). Take a point $x = x_0$ near 0 and two linearly independent solutions u_1 and u_2 of (1) at x_0. Finally take a loop C around 0 starting and ending at x_0 so that no other singularities of (1) lie inside of C. The conjugacy class of the circuit matrix $M(C)$ in $GL(2,\mathbb{C})$ is called <u>the local monodromy of (1) at</u> $x = 0$. By using a linear change on u_1 and u_2, we can assume that the circuit matrix $M(C)$ has the following Jordan's normal form:

$$M(C) = \begin{pmatrix} e^{2\pi i a} & 0 \\ 0 & e^{2\pi i b} \end{pmatrix} \quad \text{or} \quad \begin{pmatrix} e^{2\pi i a} & 1 \\ 0 & e^{2\pi i b} \end{pmatrix}$$

Put

$$N(x) = \begin{pmatrix} x^a & 0 \\ 0 & x^b \end{pmatrix} \quad \text{or} \quad \begin{pmatrix} x^a & x^a \log x \\ 0 & x^a \end{pmatrix}$$

according to the normal form of $M(C)$, respectively. Then $w = N(x)^{-1} {}^t(u_1, u_2)$ is a single valued vector.

The point $x = 0$ is called <u>a regular singular point</u> of (1) if the vector w has at most poles. Otherwise, namely when one of the components of the single valued vector w has an essential singurality, it is said to be <u>irregular</u>.

A characterization of regular singularity is given by the following <u>theorem of Fuchs</u>.

<u>Theorem</u>: The equation (1) is regular singular at $x = \xi$ if and only if $(x - \xi)p(x)$ and $(x - \xi)^2 q(x)$ are holomorphic at $x = \xi$.

<u>Proof</u>: We can assume $\xi = 0$. We prove the "only if" part first. Let x_0 be a point sufficiently near to 0 and let

u_1 and u_2 be linearly independent functions around x_0, that is to say, the Wronskian $W(u_1, u_2)$ is not identically zero. The equation with the solution space $\mathbb{C}u_1 + \mathbb{C}u_2$, defined around x_0, is given by

$$\det \begin{pmatrix} u & u' & u'' \\ u_1 & u_1' & u_1'' \\ u_2 & u_2' & u_2'' \end{pmatrix} = 0$$

If u_1 and u_2 are linearly independent solutions of (1) at x_0, then this equation must coincide with (1), which is defined also around the origin. So the coefficients p and q of (1) are given by

$$-p = \det \begin{pmatrix} u_1 & u_1'' \\ u_2 & u_2'' \end{pmatrix} \bigg/ \det \begin{pmatrix} u_1 & u_1' \\ u_2 & u_2' \end{pmatrix}$$

$$q = \det \begin{pmatrix} u_1' & u_1'' \\ u_2' & u_2'' \end{pmatrix} \bigg/ \det \begin{pmatrix} u_1 & u_1' \\ u_2 & u_2' \end{pmatrix}$$

On the other hand, under the assumption that (1) is regular singular at the origin, u_1 and u_2 have the expression ${}^t(u_1, u_2) = N(x)w$ where $N(x)$ is the matrix defined at the beginning of this section and w is a meromorphic vector around 0. Now by using the expression of u_1 and u_2, we can see the local behaviour of p and q around $x = 0$ to finish the proof. The other direction of the proof is given in the end of the next paragraph.

<u>Remark</u> : There is no analogously effective theorem for systems of the form (M). The following is only one useful sufficient condition of regular singularity :

"If a system has the form

$$x \frac{dv}{dx} = A(x)v$$

where $A(x)$ is a matrix holomorphic at 0 and v is an unknown vector, then the singularity at $x = 0$ is regular singular."

It is true that the system $x^p dv/dx = A(x)v$ (p is a positive integer and $A(x)$ is as above) has a regular singularity at $x = 0$ if and only if there exists a meromorphic transformation $v = T(x)w$ such that the system is transformed into $xdw/dx = B(x)$ with $B(x)$ holomorphic at 0. But it is difficult, in general, to find the transformation $T(x)$.

§ 2.5 The Frobenius Method

We shall construct local solutions of the differential equation (1) at a regular singular point. Consider the following operator

$$L = x^2 \frac{d^2}{dx^2} + xp^*(x) \frac{d}{dx} + q^*(x)$$

where p^* and q^* are holomorphic functions with the expansion

$$p^*(x) = \sum_{j=0}^{\infty} p_j x^j \quad \text{and} \quad q^*(x) = \sum_{j=0}^{\infty} q_j x^j.$$

Put

$$u = x^s \sum_{j=0}^{\infty} c_j x^j \qquad (c_0 = 1)$$

and calculate Lu:

$$Lu = \sum_{n \geq 0} \{(n+s)(n+s-1)c_n + \sum_{i+j=n}((i+s)c_i p_j + c_i q_j)\} x^{n+s}$$

$$= \sum_{n \geq 0} \{((n+s)(n+s-1) + (n+s)p_0 + q_0)c_n + R_n\} x^{n+s}$$

where

$R_0 = 0$, and for $n > 0$,

$$R_n = R_n(c_1, \ldots, c_{n-1}, s) = \sum_{i+j=n, i \neq n} \{(i+s)c_i p_j + c_i q_j\}.$$

If we put

$$f(s) = s(s-1) + sp_0 + q_0$$

then we see that $Lu = 0$ if and only if

$$(\#)_n : \quad f(s+n)c_n + R_n = 0$$

holds for all $n = 0, 1, 2, \ldots$.

Definition: The second order algebraic equation $(\#)_0 : f(s) = 0$ is called <u>the characteristic equation</u> (or <u>the indicial equation</u>) of the equation $Lu = 0$ at $x = 0$ and the roots of the equation are called <u>the characteristic exponents</u>.

Now consider s as a parameter, determine the coefficients $c_n = c_n(s)$ ($c_0 = 1$) by the equations $(\#)_n$ $n = 1, 2, \ldots$; and put

$$u(s,x) = x^s \sum_{i \geq 0} c_i(s) x^i.$$

Let s_1 and s_2 be the solutions of $f(s) = 0$. We can assume that $\operatorname{Re} s_1 \leq \operatorname{Re} s_2$. Then $u(s_2, x)$ is a solution of our differential equation, because we have $f(s+n) \neq 0$ for all $n \geq 1$.

If $s_2 - s_1 \neq 0, 1, 2, \ldots$, then by the same reason, $u(s_1, x)$ is another solution, which is linearly independent of $u(s_2, x)$ because of the expressions $u(s_j, x) = x^{s_j}(1 + \cdots)$ ($j = 1, 2$). This is also true when $s_2 - s_1$ is a positive integer, say m, and if R_m happens to be zero. In this case we can solve $(\#)_n$ for $s = s_1$ for all $n \geq 1$ (choose c_m arbitrarily). Otherwise, differentiate

$$Lu(s,x) = x^s f(s)$$

by s and put $s = s_2$, then we get

$$\frac{\partial}{\partial s} Lu(s,x) \big|_{s=s_2} = x^{s_2} f'(s_2)$$

In the case $s_1 = s_2$, namely when s_2 is a double root of f, since we have $f'(s_2) = 0$, the following expression

$$\frac{\partial}{\partial s} u(s,x)\Big|_{s=s_2} = u(s_2,x)\log x + x^{s_2}\sum_{j\geq 0} c'_j(s_2) x^j$$

is a solution of $Lu = 0$, because L does not depend on s and so L and $\partial/\partial s$ are commutative. It should be mentioned that since f and R_n are holomorphic with respect to s, it is also the case for the c_n's.

Finally consider the remaining case: $s_2 - s_1 = m \in \mathbb{Z}$, $m \neq 0$ and $R_m \neq 0$. Put

$$u^* = x^{s_1} \sum_{j \geq 0} c_j(s_1) x^j$$

where the c_j's ($j < m$) are determined ($c_0 = 1$) by $(\#)_j$, while c_m is arbitrarily fixed and the c_j's ($j > m$) are determined again by $(\#)_j$. Then the series u^* satisfies the following:

$$Lu^* = R_m(c_1,\cdots,c_{m-1},s_1) x^{s_1+m}$$

Since $s_1 + m = s_2$, we know that a suitable linear combination of u^* and $\frac{\partial}{\partial s} u(s,x)\Big|_{s=s_2}$, say

$$f'(s_2) u^* - R_m \frac{\partial}{\partial s} u(s,x)\Big|_{s=s_2}$$

is a solution.

So we have constructed two linearly independent solutions of $Lu = 0$ in all cases. In order to finish the proof of the theorem of Fuchs (§ 2.4) we have only to rearrange the two linearly independent solutions obtained above in a column vector and to decompose it into the product of a matrix, whose entries are root functions and logarithm function, and a vector whose entries are power series in x.

(i) If $s_2 - s_1 \neq 0, 1, 2, \cdots$ or (ii) if $s_2 - s_1 = m \neq 0$

and $R_m = 0$ (in this case the singularity is said to be non-logarithmic because no logarithm function appears in the following expressions), then the two solutions are given by

$$\begin{pmatrix} u(s_1,x) \\ u(s_2,x) \end{pmatrix} = \begin{pmatrix} x^{s_1} & 0 \\ 0 & x^{s_2} \end{pmatrix} \begin{pmatrix} \sum_{j \geq 0} c_j(s_1) x^j \\ \sum_{j \geq 0} c_j(s_2) x^j \end{pmatrix}$$

(ii) If $s_2 = s_1$, then the two solutions are given by

$$\begin{pmatrix} \frac{\partial}{\partial s} u(s,x)|_{s=s_2} \\ u(s_2,x) \end{pmatrix} = \begin{pmatrix} x^{s_2} & x^{s_2} \log x \\ 0 & x^{s_2} \end{pmatrix} \begin{pmatrix} \sum_{j \geq 0} c'_j(s_2) x^j \\ \sum_{j \geq 0} c_j(s_2) x^j \end{pmatrix}$$

(iii) If $s_2 - s_1 = m \neq 0$ and $R_m \neq 0$, then the two solutions are given by

$$\begin{pmatrix} R_m u^* - f'(s_2) \frac{\partial}{\partial s} u(s,x)|_{s=s_2} \\ u(s_2,x) \end{pmatrix} = \begin{pmatrix} x^{s_1} & x^{s_1} \log x \\ 0 & x^{s_1} \end{pmatrix}$$

$$\times \begin{pmatrix} R_m \sum_{j \geq 0} c_j(s_1) x^j - f'(s_2) x^{s_2-s_1} \sum_{j \geq 0} c'_j(s_2) x^j \\ \sum_{j \geq 0} c_j(s_2) x^j \end{pmatrix}$$

Remarks : 1) All the series that appeared in the previous calculation converge. The easy proof of this fact can be found in each standard textbook. Many sufficient conditions for differential (and functional) equations are known which guarantee the convergence of formal solutions. Recently R. Gérard (in his contribution in [G-O]) proved an elegant theorem which includes all convergence theorems known so far.

2) A non-logarithmic singularity with the exponents $s_1 = 0$, $s_2 = 1$ is nothing but a non-singular point.

3) By the definition of the characteristic equation $f(s) = 0$, we see that

$$s_1 + s_2 = 1 - p_0$$

and that p_0 is the residue of the coefficient $p(x) = p^*(x)/x$ at $x = 0$.

4) If the coefficients p_j and q_j as well as s_1 and s_2 are real and $s_2 - s_1$ is not an integer then choosing branches of x^{s_j} suitably, the solutions $u(s_j,x)$ ($j = 1,2$) are real valued when x varies on a small interval $(0,\varepsilon)$ ($\varepsilon > 0$).

Up to now we have studied only local properties. We now turn to the global situation.

§ 2.6 Fuchsian Equations

The equation (1) is called **Fuchsian** if all singularities in the projective line $\mathbb{P}_1 = \mathbb{C} \cup \{\infty\}$ are regular singular.

The following table of the singular points and the exponents is called the **Riemann scheme**.

$$\begin{pmatrix} x = x_1 & \cdots & x_m & x_{m+1} = \infty \\ s_1^{(1)} & \cdots & s_1^{(m)} & s_1^{(m+1)} \\ s_2^{(1)} & \cdots & s_2^{(m)} & s_2^{(m+1)} \end{pmatrix}$$

<u>Proposition</u> (A characterization of Fuchsian equations): The equation (1) is Fuchsisian with regular singularity at $S = \{x_1, \cdots, x_m, x_{m+1} = \infty\}$ if and only if the coefficients p and q have the following form:

$$p = \sum_{j=1}^{m} \frac{a_j}{x - x_j} \qquad (a_j \in \mathbb{C})$$

$$= \{\text{a polynomial of degree at most } (m-1)\} / \prod_{j=1}^{m}(x - x_j)$$

$$q = \{\text{a polynomial of degree at most } 2(m-1)\} / \prod_{j=1}^{m}(x - x_j)^2$$

<u>Proof</u>: Apply the Fuchs theorem to the equation (2) at infinity.

Proposition (The Fuchs Relation): The sum of all exponents of (1) depends only on the number of singular points:

$$\sum_{j=1}^{m+1} \sum_{i=1}^{2} s_i^{(j)} = m - 1.$$

Proof: At a finite point $x = x_j$, we know (Remark 3 in § 2.5) that $s_1^{(j)} + s_2^{(j)} = 1 - a_j$. At infinity, with $t = 1/x$, equation (1) changes into equation (2). Thus we have

$$s_1^{(m+1)} + s_2^{(m+1)} = 1 - \text{Res}_{t=0}\{\tfrac{2}{t} - \tfrac{1}{t^2}p(\tfrac{1}{t})\}$$

$$= 1 - 2 + \sum_{j=1}^{m} a_j,$$

which completes the proof.

Remark: 1) The Fuchsian relation can be considered as the residue theorem applied to the 1-form dW/W, where W is the Wronskian of (1). Indeed since the logarithmic derivative of W is expressed in the finite plane by $dW/W = -p(x)dx$ and since at infinity we have $\text{Res}_{x=\infty} \frac{dW}{W} = -\tfrac{1}{t^2}p(\tfrac{1}{t})\big|_{t=0}$, the following identity holds:

$$\sum_{x \in P_1} \text{Res}_x \frac{dW}{W}$$

$$= \{\sum_{j=1}^{m}(s_1^{(j)} + s_2^{(j)} - 1)\} + \{s_1^{(m+1)} + s_2^{(m+1)} + 1\}.$$

With the previously introduced terminology, we are now able to say that the HGDE is a second order Fuchsian equation with regular singular points at 0, 1 and ∞; that its Riemann scheme is given by

$$\begin{Bmatrix} x = 0 & x = 1 & x = \infty \\ 0 & 0 & a \\ 1-c & c-a-b & b \end{Bmatrix}$$

and that the sum of its characteristic exponents is 1.

Remarks: 2) A second order Fuchsian equation with only two singular points is trivial; it is solved by x^s or by log x. So the HGDE is the simplest non-trivial Fuchsian equation.

3) For any three distinct points $\{x_1, x_2, x_3\}$ in $\mathbf{P}_1$ and any system of complex numbers $\{s_i^{(j)}\}$ ($i = 1,2$; $j = 1,2,3$) satisfying $\sum_{i,j} s_i^{(j)} = 1$, there is a uniquely determined second order Fuchsian differential equation of the form (1) which has the following Riemann scheme.

$$\begin{Bmatrix} x = x_1 & x = x_2 & x = x_{m+1} \\ s_1^{(1)} & s_1^{(2)} & s_1^{(m+1)} \\ s_2^{(1)} & s_2^{(2)} & s_2^{(m+1)} \end{Bmatrix}$$

If we use the projective transformation on the variable x sending $\{x_1, x_2, x_3\}$ to $\{0, 1, \infty\}$ and if we next change the unknown u into $x^{-s_1^{(1)}} (1-s)^{-s_2^{(2)}} u$ then the Riemann scheme becomes

$$\begin{Bmatrix} x = 0 & x = 1 & x = \infty \\ 0 & 0 & \sum_{j=1}^{3} s_1^{(j)} \\ s_2^{(1)} - s_1^{(1)} & s_2^{(2)} - s_1^{(2)} & \sum_{j=1}^{3} s_2^{(j)} \end{Bmatrix}$$

So the Fuchsian equation we started with has been reduced to the HGDE.

Chapter 3 The Riemann and Riemann-Hilbert Problems

In this chapter the Riemann and the Riemann-Hilbert problems are stated. The first problem is so easy to solve that one might say it is almost obvious. Nevertheless it is discussed here in order to prepare for the exposition of the same question in several variables studied in Chapter 9 which is by no means easy. The second problem was first solved on general Riemann surfaces by Röhrl [Röh]. Here we present an improved version due to Ohtski [Oht 2]. The proof of Ohtski's result uses some of the language of algebraic geometry with which we assume the reader to be familiar. The question posed by Riemann-Hilbert gave rise to the theory of iso-monodromic deformation of a linear differential equation, which is presently one of the most active topics in the theory of differential equations.

§ 3.1 Statement of the Problems

In order to state our two problems let X be the complex projective line P_1 and $S = \{x_1, \ldots, x_{m+1}\}$ a finite subset of $m+1$ points of X.
The Riemann problem: Let $s_i^{(j)}$ ($j = 1, \ldots, m+1$; $i = 1,2$) be complex numbers. Find a Fuchsian differential equation (1) which has regular singular points at S with the following Riemann scheme :

$$\begin{Bmatrix} x = x_1 & \cdots & x = x_{m+1} \\ s_1^{(1)} & \cdots & s_1^{(m+1)} \\ s_2^{(1)} & \cdots & s_2^{(m+1)} \end{Bmatrix}$$

The Riemann-Hilbert Problem: Let $\rho : \Pi_1(X - S) \to GL(2,\mathbb{C})$ be a representation. Find a Fuchsian equation (1) which has ρ as its monodromy representation.

Remark: The Riemann problem starts from <u>local</u> data in order to end up with a <u>global</u> object, while the Riemann-Hilbert problem involves getting the <u>global</u> object from <u>global</u> initial information.

§ 3.2 An Observation

Consider a Fuchsian equation of the form (1) in Chapter 2 with regular singularities on S. Let E_0 be the number of characteristic exponents which satisfy the Fuchs relation, let E be the number of coefficients of $p(x)$ and $q(x)$ and let M be the number of parameters of homomorphisms modulo conjugacy of $\Pi_1(X - S)$ to $GL(2,\mathbb{C})$. Since the number of characteristic exponents $\{s_i^{(j)}\}$ is $2(m + 1)$, we have

$$E_0 = 2(m + 1) - 1 = 2m + 1.$$

As we know the denominators and the degrees of the numerators of the rational functions p and q (§ 2.6 Proposition), we have

$$E = (m - 1) + 1 + 2(m - 1) + 1 = 3m - 1.$$

The group $\Pi_1(X - S)$ is the free group generated by m elements; for instance by loops around $x_1, \ldots, x_m$. So taking account of conjugacy, we see that

$$M = 4m - (4 - 1) = 4m - 3.$$

Notice that if $m = 2$ (this is the case of the HGDE) then $E_0 = E = M = 5$, while if $m \geq 3$ then $E_0 < E < M$.

§ 3.3 The Solution of the Riemann Problem

Prescribing the system of local data $\{s_i^{(j)}\}$ ($i = 1, 2$; $j = 1, \ldots, m+1$) amounts to requiring certain inhomogeneous linear equalities for the E coefficients of p and q. It is easy to check that this system has a solution if and only if the given $\{s_i^{(j)}\}$ satisfies the Fuchs relation. Therefore we conclude that the Riemann problem is solvable if and only if the local datum satisfies the Fuchsian relation and that the solutions form an affine space of dimension $E - E_0 = m - 2$.

<u>Definition</u>: The coefficients of a differential equation not depending on local data are called <u>accessory parameters</u>.

<u>Remark</u>: Fuchsian equations which are free of accessory parameters are of great interest. Fuchsian ordinary differential equations of order two have this property only if $\#S \leq 3$. But there are many equations of order greater than two having the property. These have been tabulated by K. Okubo ([Oku]). See also § 9.2 where higher dimensional cases are treated.

§ 3.4 Apparent singularities

Let us now turn to the Riemann-Hilbert problem. If one insists on requiring that the Fuchsian equation (1) has singularities only on S, the problem has in general no solution, because we know that if $m \geq 3$ then $M > E$. (It is also known that even if $m = 2$ the problem is not always solvable). Thus in order to solve it we must consider differential equations of the form (1) admitting perhaps regular singularities besides those of S.

Definition: A point $x = x_0$ is called an __apparent singularity__ of (1) if x_0 is a singularity of (1) and the local monodromy of (1) at x_0 is trivial.

Remark: The following statements are equivalent. 0) A point x_0 is an apparent singularity of (1). 1) A point x_0 is a singularity of (1) and there exist two linearly independent holomorphic solutions around x_0. 2) A point x_0 is a non-logarithmic singularity of (1) such that the two characteristic exponents are non-negative integers. 3) There exist two linearly independent holomorphic solutions of (1) around x_0 whose Wronskian vanishes at x_0.

Consider now a Fuchsian equation (1) with regular singularities only on S. If we allow an extra singularity ξ outside S then E increases by 3. But the requirment that ξ be an apparent singularity imposes three conditions on the coefficients of p and q namely that the two characteristic exponents be integers and that ξ be a non-logarithmic singularity. Thus the number of parameters of p and q does not change by allowing a fixed extra apparent singularty. But if we allow this singularity to vary the number of parameters increases by 1. On the other hand whether ξ is allowed to move or not the number M remains unchanged, since the local monodromy at x_0 is trivial. Therefore in order to solve the Riemann-Hilbert problem for a general representation ρ we must allow M - E apparent singular points outside S. The next paragraph gives sufficient conditions for the solution of the problem.

§ 3.5 A Solution of the Riemann-Hilbert Problem

Let

(3) $\quad u^{(n)} + p_{n-1} u^{(n-1)} + \cdots + p_1 u' + p_0 u = 0$

be a Fuchsian differential equation of order n with $m+1$ regular singular points $x_1, \ldots, x_m$ and x_{m+1}, and let $S = \{x_1, \ldots, x_{m+1}\}$. In this case the number E of parameters of the coefficients p_j ($j = 0, \ldots, n-1$) and the number M of parameters of conjugate classes of homomorphisms of $\Pi_1(X - S)$ to $GL(n,\mathbb{C})$ are given by

$$E = (m-1)+1 + 2(m-1)+1 + \cdots + n(m-1)+1$$

$$= n\{n(m-1) + m + 1\}/2$$

and $\quad M = n^2 m - (n^2 - 1)$.

A homomorphism ρ of a group G into $GL(n,\mathbb{C})$ is called <u>irreducible</u> if there is no non-trivial linear subspace of $\mathbb{C}^n$ invariant under the action of the image group $\rho(G)$.

<u>Theorem</u> ([Oht 2]): Let X and S be as above and let ρ be a homomorphism of $\Pi_1(X - S)$ to $GL(n,\mathbb{C})$. Assume that ρ is irreducible and assume further that there exists an index j such that the circuit matrix around x_j has at least one Jordan-block of size one. Then there exists a Fuchsian differential equation (3) on X with regular singularities on $S \cup S'$ whose monodromy representation is ρ, where S' consists of at most $M - E$ apparent singularities.

<u>Remarks</u> 1) In the course of the following proof of the theorem, it is assumed that the reader is familiar with the foundation of algebraic geometry. (Deligne's lecture note [Del] is enough.) Since we never use this proof in this book, if the reader is not familiar with such a subject, it is advisable to skip the proof.
 2) The theorem also holds word for word on any compact Riemann surface of genus g. In order to make the proof remain valid in this more general case we introduced the genus g which must be taken to zero in the present context.

Proof: The local system $\underline{V}'$ associated with the given ρ (see [Del]) determines on $X - S$ a vector bundle (locally constant sheaf) $\underline{V}' = V' \otimes_{\mathbb{C}} O_{X-S}$ and a holomorphic integrable connection D' on $\underline{V}'$ (i.e. a $\mathbb{C}$-linear map over $X - S$: $\underline{V}' \to \Omega^1_{X-S} \otimes_{O_{X-S}} \underline{V}'$ satisfying the Leibniz formula $D'(fv) = df \otimes v + fD'v$ for $f \in O_{X-S}$ and $v \in \underline{V}'$) such that for all $v \in V' \subset \underline{V}'$ one has $D'v = 0$. The last relation means that V' is the locally constant sheaf of horizontal sections of the conection D' in $\underline{V}'$. Here O_{X-S} and Ω^1_{X-S} stand for sheaves of germs of holomorphic functions and holomorphic 1-forms on $X - S$, respectively. We now proceed to extend the holomorphic vector bundle with connection $(\underline{V}',D')$ in order to obtain a holomorphic vector bundle V on X with a meromorphic connection D having simple poles at points of S. In order to carry out the construction we work with the monodromy around each point $x_j \in S$. Let A_j be the circuit matrix around x_j and B_j be a matrix such that

$$A_j = \exp(-2\pi i B_j).$$

For an open contractible neighborhood U of x_j in X not containing other points of S define the connection D_U on O_U^n by setting

$$D_U v := dv + x^{-1} B_j dx \otimes v \qquad (v \in O_U^n)$$

where x is a local coordinate around x_j, and check that $D_U v = 0$ if and only if $v = x^{-B_j}$. Notice that (O_U^n, D_U) can be patched to $(\underline{V}',D')$ and do this for all j ($j = 1, \ldots, m+1$). The result is the extension $(\underline{V},D)$. Using this construction one sees that the first Chern class of $\underline{V}$ or equivalently the Chern number $c_1(\underline{V})$ of the determinant bundle $\Lambda^n \underline{V}$ of $\underline{V}$ is given by

$$c_1(\underline{V}) = - \sum_{j \in S} \text{tr}(B_j).$$

Since by assumption there is an index j such that the Jordan normal form of A_j can be written as

$$\begin{pmatrix} a & 0 & \cdots & 0 \\ 0 & & & \\ & & \star & \\ 0 & & & \end{pmatrix}$$

the matrix B_j can be written as

$$\begin{pmatrix} \log a + 2\pi i k & 0 & \cdots & 0 \\ 0 & & & \\ & & \star & \\ 0 & & & \end{pmatrix} \qquad (k \in \mathbf{Z})$$

This shows that the extension $(\underline{V}',D)$ is not unique, since the Chern number $c_1(\underline{V})$ can be chosen arbitrarily. This remark will soon prove to be important. We have constructed a connection D on $\underline{V}$ such that n linearly independent horizontal sections say, $v_1, \ldots, v_n$ have the prescribed monodromy behaviour. We now construct a differential equation (3) such that n linearly independent solutions have the same monodromy behaviour. In order to do so assume for a moment that there exists a non-trivial global holomorphic section v^* of the dual bundle $\underline{V}^*$ of $\underline{V}$. Now choose a system of $\mathbf{C}$ bases $v_1, \ldots, v_n$ of V' and consider the following equation in the unknown u:

$$\det \begin{pmatrix} (v^*,v_1) & \cdots & u \\ (v^*,v_1)' & \cdots & u' \\ & & \\ (v^*,v_1)^{(n)} & \cdots & u^{(n)} \end{pmatrix} = 0$$

This is the desired equation (3) with linearly independent solutions $(v^*,v_1), \ldots, (v^*,v_n)$. Indeed since by assumption the representation ρ is irreducible, we have an isomorphism

$$\underline{V}' \supset V' \xrightarrow{\cong} v^*(V') \subset \mathcal{O}_{X-S}$$

and so the Wronskian W is non-trivial.

It is easy to see that W has poles at each point x_j of order at most $n(n-1)/2$. The set of zeros of W is the the set of apparent singularities of this equation (see § 3.4 Remark). Let us estimate the number of zeros of W. Since one can check that W is a meromorphic section of $(\Lambda^n \underline{V}^*) \otimes \Omega^{n(n-1)/2}$, we have

$$\#\{\text{zeros of } W\} - \#\{\text{poles of } W\} = c_1(\underline{V}^*) + (2g-2)n(n-1)/2$$

where # denotes cardinarity of sets. As we saw above, we have the estimate

$$\#\{\text{poles of } W\} \leq (m+1)n(n-1)/2.$$

When $\underline{V}^*$ admits a nontrivial global section, We want to have $c_1(\underline{V}^*)$ as small as possible. The Hirzebruch-Riemann-Roch theorem tells us that

$$\dim H^0(X, O(\underline{V}^*)) \geq c_1(\underline{V}^*) + n(1-g).$$

where $H^0(X, O(\underline{V}^*))$ denotes the space of global holomorphic sections of $\underline{V}^*$. So the most economic way is to choose $\underline{V}$ so that

$$c_1(\underline{V}^*) = 1 - n(1 - g).$$

Eventually we get the following estimate

$$\#\{\text{poles of } W\} = \#\{\text{accessory parameters}\}$$

$$\leq 1 - n(1-g) + (m+1+2g-2)n(n-1)/2$$

$$= M - E.$$

This concludes the proof of Ohtski's theorem.

§ 3.6 Isomonodromic Deformation

Knowing that the Riemann-Hilbert problem has a solution, it is natural to examine the structure of the space of solutions namely to give a description of the space of equations with a fixed monodromy group. This problem is sometimes called the Fuchs problem or more precisely the problem of isomonodromic deformations. In this paragraph, following Okamoto ([Okm *]), we formulate the problem and give a solution without proof. If the reader becomes interested in the theory please consult the original papers. Consider the differential equation

(1) $u'' + p(x)u' + q(x)u = 0$

with $N+2$ non-apparent regular singularities in S and N (= $M - E$) apparent singularities in S' where we set

$$S = \{0, 1, t_1, \ldots, t_N, \infty\} \quad \text{and} \quad S' = \{\lambda_1, \ldots, \lambda_N\}.$$

Changing the dependent variable by multiplying with a function of x, we can assume without loss of generality that one of the exponents at each finite singular point is zero. In order to avoid complications, let us restrict to the case when the characteristic exponents at the λ_k's are 0 and 2 (cf.[Kim 2]). Consider the Riemann scheme

$$\left\{ \begin{array}{ccccc} x = 0 & x = 1 & x = t_j & x = \lambda_k & x = \infty \\ 0 & 0 & 0 & 0 & \rho_\infty \\ \kappa_0 & \kappa_1 & \theta_j & 2 & \rho_\infty + \kappa_\infty \end{array} \right\}$$

with the Fuchs relation

$$\kappa_0 + \kappa_1 + \sum \theta_j + \kappa_\infty + 2\rho_\infty + 2N = 2N + 2 - 1.$$

Equation (1) with this Riemann scheme has $2N$ accessory parameters (§ 3.1). Indeed with the help of $2N$ newly introduced constants H_j, μ_k ($j,k = 1,\ldots,N$) the coefficients $p(x)$ and $q(x)$ can be expressed as follows:

$$p(x) = \frac{1-\kappa_0}{x} + \frac{1-\kappa_1}{x-1} + \sum \frac{1-\theta_j}{x-t_j} - \sum \frac{1}{x-\lambda_k}$$

$$q(x) = \frac{1-\kappa}{x(x-1)} - \sum \frac{t_j(t_j-1)H_j}{x(x-1)(x-t_j)} + \sum \frac{\lambda_k(\lambda_k-1)\mu_k}{x(x-1)(x-\lambda_k)}$$

where $\kappa = \rho_\infty(\rho_\infty + \kappa_\infty)$. Since λ_k is an apparent singularity with exponents $\{0, 2\}$, the non-logarithmic condition (§ 2.5) tells us that each H_j is a rational function of the λ_k's, μ_k's and t_i's. Indeed it is explicitly given as follows

$$H_j = M_j \sum_k M^{k,j}\{\mu_k^2 - (\frac{\kappa_0}{\lambda_k} + \frac{\kappa_1}{\lambda_k-1} + \sum_i \frac{\theta_i - \delta_{ij}}{\lambda_k - t_i}\mu_k) + \frac{\kappa}{\lambda_k(\lambda_k-1)}\}$$

where the δ_{ij}'s are Kronecker symbols and

$$M_j = -\frac{\Lambda(t_j)}{T'(t_j)}, \qquad M^{k,j} = \frac{T(\lambda_k)}{\Lambda'(\lambda_k)(\lambda-t_j)}.$$

$$T(x) = x(x-1)\Pi_j(x-t_j), \qquad \Lambda(x) = \Pi_k(x-\lambda_k),$$

$$T'(x) = \frac{dT(x)}{dx}, \qquad \Lambda'(x) = \frac{d\Lambda(x)}{dx}.$$

Let us fix the monodromy representation of (1) (this implies in particular that the exponents κ_0, κ_1, θ_j, ρ_∞ and κ_∞ are fixed) and let the λ_k's, μ_j's and the t_i's vary. Our problem now turns out to be the problem of describing the t_i-dependence of the λ_k's and the μ_j's, which is solved by the following theorem.

<u>Theorem</u> (Okamoto): The monodromy group of (1) is unchanged if and only if, considered as functions of the t_i's, the λ_k's and the μ_j's satisfy the "completely integrable Hamilton system"

$$G_N : \quad \frac{\partial \lambda_k}{\partial t_j} = \frac{\partial H_j}{\partial \mu_k}, \quad \frac{\partial \mu_k}{\partial t_j} = -\frac{\partial H_j}{\partial \lambda_k} \quad (1 \leq j,k \leq N)$$

Remarks: 1) The system G_1 is equivalent to the Painlevé equation. If we eliminate μ_1 then it turns out to be the famous second order (non-linear) ordinary differential equation for $\lambda_1(t_1)$ called the sixth Painleve equation.

$$\lambda'' = \frac{1}{2}(\frac{1}{\lambda} + \frac{1}{\lambda-1} + \frac{1}{\lambda-t})(\lambda')^2 - (\frac{1}{t} + \frac{1}{t-1} + \frac{1}{\lambda-t})\lambda'$$
$$+ \frac{\lambda(\lambda-1)(\lambda-t)}{t^2(t-1)^2}\{\alpha + \beta\frac{t}{\lambda^2} + \gamma\frac{t-1}{(\lambda-1)^2} + \delta\frac{t(t-1)}{(\lambda-t)^2}\}$$

where $\lambda = \lambda_1$, $t = t_1$, $\lambda' = \frac{d\lambda}{dt}$, $2\alpha = \kappa_\infty^2$, $2\beta = -\kappa_0^2$, $2\gamma = \kappa_1^2$ and $2\delta = 1 - \theta_1^2$.

2) The systems G_N ($N \geq 2$) are called the Garnier systems. We shall discuss the system G_N again in § 6.6.

Chapter 4 Schwarzian Derivatives I

The Schwarzian derivative introduced in this chapter is the main tool which enables one to connect the theory of orbifolds with the theory of linear differential equations. We give extensive computations since they are easy and enlightening, especially in Chapter 10, where the notion is generalized to several variables.

§ 4.1 Definitions and Properties

The Schwarzian derivative $\{z;x\}$ of a non-constant smooth function z of x with respect to x is defined as follows

$$\{z;x\} = \left(\frac{dz}{dx}\right)^{\frac{1}{2}} \left(\frac{d^2}{dx^2}\right)\left(\frac{dz}{dx}\right)^{-\frac{1}{2}}$$

$$= \frac{3\left(\frac{d^2z}{dx^2}\right)^2 - 2\frac{dz}{dx}\frac{d^3z}{dx^3}}{4\left(\frac{dz}{dx}\right)^2}$$

The second expression shows that it is a rational function of the derivatives z', z'' and z''' of z with respect to x, whose denominator is $(z')^2$.

<u>Proposition</u>: The Schwarzian derivative has the following properties:

1) (PGL(2,$\mathbb{C}$)-invariance)

$$\left\{\frac{az+b}{cz+d}; x\right\} = \{z;x\} \quad \text{for all} \quad \begin{pmatrix} a & b \\ c & d \end{pmatrix} \in GL(2,\mathbb{C})$$

2) (the connection formula) If x is a non-constant function of y, then

$$\{z;y\} = \{x;y\} + \{z;x\}\left(\frac{dx}{dy}\right)^2$$

Remarks: 1) By the properties 1) and 2) one sees that $\{z;x\} = 0$ if and only if the map $x \mapsto z(x)$ is projective linear i.e. $z(x) \in PGL(2,\mathbb{C})$.

2) The connection formula can be also expressed as the relation between the three quadratic differentials:

$$\{z;y\}(dy)^2 = \{x;y\}(dy)^2 + \{z;x\}(dx)^2.$$

Although the proposition can be checked by direct computation, we prefer to prove it by making use of the relations between Schwarzian derivatives and differential equations in order to make the meaning of the proposition clear. This is done in the following two sections.

A function $z(x)$ is said to be $PGL(2,\mathbb{C})$-<u>multivalued</u> if any two branches of $z(x)$ are projectively related. Using this terminology, assersion (1) tells that if a function $z = z(x)$ is $PGL(2)$-multivalued then the Schwarzian derivative $\{z;x\}$ is single valued.

§ 4.2   Relations with Differential Equations

<u>Proposition</u>: Let $z(x)$ be a non-constant $PGL(2)$-multivalued function and put $q(x) := -\{z;x\}$. Then there are two linealy independent solutions u_0 and u_1 of the equation

$$\frac{d^2u}{dx^2} + q(x)u = 0$$

with single valued coefficients such that $z(x) = \dfrac{u_0(x)}{u_1(x)}$.

Proof: Put

$$u_0 = \left(\frac{dz}{dx}\right)^{-\frac{1}{2}} \quad \text{and} \quad u_1 = zu_0$$

and consider the linear differential equation with $\mathbb{C}u_0 + \mathbb{C}u_1$ as its solution space, which is given by

$$(*) \quad \det \begin{pmatrix} u_0 & u_0' & u_0'' \\ u_1 & u_1' & u_1'' \\ u & u' & u'' \end{pmatrix} = 0$$

Now consider a transformation

$$z \mapsto \frac{az + b}{cz + d} \quad \text{with} \quad ad - bc = 1.$$

Since $\frac{dz}{dx}$ changes into $(cz + d)^{-2} \frac{dz}{dx}$, u_0 and u_1 change as follows

$$u_0 \to (cz + d)u_0 = cu_1 + du_0$$

$$u_1 \to \frac{az + b}{cz + d}(cz + d)u_0 = au_1 + bu_0$$

This shows that if z changes projectively then u_0 and u_1 change linearly

$$\begin{pmatrix} u_0 \\ u_1 \end{pmatrix} \to \begin{pmatrix} d & c \\ b & a \end{pmatrix} \begin{pmatrix} u_0 \\ u_1 \end{pmatrix}$$

and that, since $\begin{pmatrix} d & c \\ b & a \end{pmatrix} \in SL(2)$, the above equation $(*)$ must have single valued coefficients. Let us compute the coefficients of $(*)$.
The coefficient of u'' is

$$\det \begin{pmatrix} u_0 & u_0' \\ u_1 & u_1' \end{pmatrix} = u_0(z'u_0 + zu_0') - u_0'zu_0$$

The coefficient of u' is

$$-\det\begin{pmatrix} u_0 & u_0'' \\ u_1 & u_1'' \end{pmatrix} = -u_0(z''u_0 + 2z'u_0' + zu_0'') + u_0 z u_0''$$

$$= -\frac{z''}{z'} - 2z'(-\frac{1}{2})(z')^{-\frac{3}{2}} z'' u_0$$

$$= 0.$$

The coefficient of u is

$$\det\begin{pmatrix} u_0' & u_0'' \\ u_1' & u_1'' \end{pmatrix} = u_0'(z''u_0 + 2z'u_0' + zu_0'') - u_0''(z'u_0 + zu_0')$$

$$= u_0'\{z''(z')^{-\frac{1}{2}} + 2z'(-\frac{1}{2})(z')^{-\frac{3}{2}} z''\}$$

$$- (z')^{\frac{1}{2}}((z')^{-\frac{1}{2}})''$$

$$= -\{z;x\}.$$

These computations show that the equation (*) is nothing but the equation

$$u'' + q(x)u = 0$$

which has the desired properties. This concludes the proof of the proposition as well as that of assertion (1) of the proposition in § 4.1.

§ 4.3 A Canonical Form

Definition: The equation

(1) $\quad u'' + pu' + qu = 0$

is said to be in (projectively) <u>canonical form</u> if $p = 0$.

Remark: It is equivalent to require that its Wronskian be constant. In this case the monodromy group is contained in $SL(2,\mathbb{C})$.

Definition: The ratio of two linearly independent solutions is called a <u>projective solution</u>. It is determined up to a transformation in $PGL(2,\mathbb{C})$. Its class is called <u>the projective solution</u>. Two differential equations with the same projective solution are said to be <u>projectively equivalent</u>.

Proposition: By replacing the unknown u by its product with a non-zero function of x, any equation of the form (1) can be transformed into a uniquely defined equation in canonical form namely into

$$u'' + (q - \frac{p'}{2} - \frac{p^2}{4})u = 0.$$

Proof: Substitute for the unknown u by putting $v = au$ where $a = a(x)$ is a non-zero function of x. The transformed equation is

$$v'' + (p + \frac{2a'}{a})v' + (q + \frac{pa'}{a} + \frac{a''}{a})v = 0.$$

Choose $a(x)$ so that $p + \frac{2a'}{a} = 0$ and use the identity

$$\frac{a''}{a} - (\frac{a'}{a})^2 = -\frac{p'}{2}$$

to complete the proof.

Remarks: 1) The Schwarzian derivative of a projective solution of the equation (1) is the coefficient of its normal form.

2) The Riemann scheme

$$\begin{pmatrix} x_1 & \cdots & x_m & x_{m+1} = \infty \\ s_1^{(1)} & \cdots & s_1^{(m)} & s_1^{(m+1)} \\ s_2^{(1)} & \cdots & s_2^{(m)} & s_2^{(m+1)} \end{pmatrix}$$

of an equation (1) in canonical form is characterised by

$$s_1^{(j)} + s_2^{(j)} = 1 \quad (j = 1, \cdots, m)$$

$$s_1^{(m+1)} + s_2^{(m+1)} = -1.$$

Proof of the connection formula (assersion (2) of the proposition in § 4.1): Let $z = z(x)$ be a non-constant function of x. We start from the equation in canonical form

$$u'' + q(x)u = 0 \quad \text{where} \quad -q(x) = \{z;x\}$$

of which projective solution is $z(x)$. Change the variable x into a new variable y so that the equation becomes

$$(\frac{dy}{dx})^2 \frac{d^2u}{dy^2} + \frac{d^2y}{dx^2} \frac{du}{dy} + q(x(y))u = 0.$$

The canonical form of this equation is

$$\frac{d^2u}{dy^2} + \{(\frac{dy}{dx})^{-2} q - \frac{1}{2}\frac{d}{dy}(\frac{d^2y}{dx^2}(\frac{dy}{dx})^{-2}) - \frac{1}{4}((\frac{dy}{dx})^{-2}(\frac{d^2y}{dx^2})^2\}u = 0.$$

Since the coefficient of this equation should be equal to $-\{z;y\}$, we have

$$-\{z;y\} = (\frac{dy}{dx})^{-2}\{q - \frac{1}{2}\frac{d^3y}{dx^3}(\frac{dy}{dx})^{-1} + \frac{3}{4}(\frac{d^2y}{dx^2})^2(\frac{dy}{dx})^{-2}\}$$

$$= (\frac{dy}{dx})^{-2}(-\{z;x\} + \{y;x\}).$$

This completes the proof of the connection formula.

§ 4.4 Local Behavior of the Schwarzian Derivative

Proposition: Let $z(x)$ be a non-constant function of x around $x = 0$ of the form $z = x^s f(x)$ or $z = \log(xf(x))$ where $f(x)$ is a non-vanishing holomorphic function. Then we have

$$x^2 \{z(x);x\}\big|_{x=0} = -\frac{1-s^2}{4} \quad \text{or} \quad -\frac{1}{4}, \text{ respectively.}$$

Proof: In the first case, we have

$$\{z;x\} = \frac{1}{4} \frac{3z'' - 2z'z'''}{z'^2}$$

$$= \frac{3(s(s-1)x^{s-2}u + \cdots)^2 - 2(sx^{s-1}u + \cdots)(s(s-1)(s-2)x^{s-3}u + \cdots)}{4(sx^{s-1}(u+s^{-1}xu'))^2}$$

$$= \frac{(3(s-1)^2 x^{2s-4} u^2 - 2(s-1)(s-2) x^{2s-4} u^2 + \cdots)}{4x^{2s-2}(u+s^{-1}xu')^2}$$

$$= \frac{1}{4x^2}(3(s-1)^2 - 2(s-1)(s-2) + \text{higher order terms})$$

$$= -\frac{1-s^2}{4x^2} + \frac{1}{x}(\text{a holomorphic function}).$$

A similar computation is also valid in the second case. It leads to

$$\{z;x\} = -\frac{1}{4x^2} + \frac{1}{x}(\text{a holomorphic function}).$$

Remark: A map $x \mapsto z = z(x)$ is biholomorphic at $x = 0$ if and only if $\{z;x\}$ is holomorphic at $x = 0$.

Proof: If z is biholomorphic at $x = 0$, then $z'(0)$ is nonzero so, by definition, $\{z;x\}$ is holomorphic. In order to prove the converse, we consider the differential equation: $u'' - \{z;x\}u = 0$. At the non-singular point $x = 0$, there are two solutions u_0 and u_1 such that u_0 and $x^{-1}u_1$ are non-vanishing holomorphic functions (§ 2.5 Remark 2). Since u_1/u_0 is biholomorphic at $x = 0$ and since z is projectively related to u_1/u_0, z is also biholomorphic at $x = 0$.

Chapter 5 The Gauss-Schwarz Theory for Hypergeometric Differential Equations

In order to have a good understanding of the Gauss-Schwarz theory, it is convenient to introduce the notions of orbifolds, their uniformizations and the uniformizing differential equations. In the simplest non-trivial case, the uniformizing differential equation will turn out to be the HGDE.

§ 5.1 Orbifolds and their Uniformizations

Let X be a complex manifold, S be a hypersurface of X, $S = \bigcup_j S_j$ be its decomposition into irreducible components, and let b_j be either infinity or an integer (≥ 2) called the <u>weight</u> attached to the corresponding S_j. The triple $\underline{X} = (X,S,b)$ is called an <u>orbifold</u> if for every point in $X - \bigcup \{S_j \mid b_j = \infty\}$ there is an open neighborhood U and a covering manifold which ramifies along $U \cap S$ with the given indices b. It is called <u>uniformizable</u> if there is a global covering manifold (called a <u>uniformization</u>) of X with the given ramification datum (S,b). Let us assume that $\underline{X}$ is uniformizable. Then there exists a "biggest" uniformization M, called the <u>universal uniformization</u>, which is uniquely determined up to isomorphism. Now following [Kat 2], we formulate these ideas, more precisely. Let

$X_0 = X - S$

$\tilde{X}_0 =$ the universal covering of X_0

$H = \Pi_1(X_0, a)$: The fundamental group of X_0 with base point $a \in X_0$.

$\mu_j \in H$: the class represented by a loop starting from a, following a path in X_0 to a point of S_j, at where S is non-singular, then going once around the hypersurface S_j and returning to a along the path used in the outward direction.

$H(\mu^b) =$ the smallest normal subgroup of H containing all the $(\mu_j)^{b_j}$'s such that $b_j < \infty$. (This does not depend on the choice of the μ_j's.)

If X' is any uniformization of the orbifold $\underline{X} = (X, S, b)$, then there exists a normal subgroup K of H including $H(\mu^b)$ such that X' turns out to be a completion of the covering X'_0 of X_0 (that is, $X' - X'_0$ is contained in a subvariety of X') corresponding to K.

<u>Note</u> : The manifold X' is uniquely determined as the manifold Y containing X'_0 having the following properties: (i) $Y - X'_0$ is contained in a subvariety of Y, (ii) there is a surjective map $p: Y \to X - \bigcup \{ S_j | b_j = \infty \}$, which is an extention of the covering map $X'_0 \to X_0$, and (iii) any fibre of p is discrete.

In particular the universal uniformization M is the covering corresponding to the group $K = H(\mu^b)$.

The following diagram shows the correspondence.

$$
\begin{array}{ccccccc}
 & & & & \tilde{X}_0 & \longleftrightarrow & 1 \\
 & & & & \downarrow & & | \\
1 & \longleftrightarrow & M & \hookleftarrow & M_0 & \longleftrightarrow & H(\mu^b) \\
| & & \downarrow & & \downarrow & & | \\
K/H(\mu^b) & \longleftrightarrow & X' & \hookleftarrow & X'_0 & \longleftrightarrow & K \\
| & & \downarrow & & \downarrow & & | \\
H/H(\mu^b) & \longleftrightarrow & X & \hookleftarrow & X_0 & \longleftrightarrow & H \\
 & & \uparrow & & \uparrow & & \\
 & & \multicolumn{3}{c}{\text{Completion.}} & &
\end{array}
$$

Galois corresp. of   Galois corresp. of
ramified coverings.   topological coverings.

The group $\Gamma = H/H(\mu^b)$ can be considered as a properly discontinuous transformation group of M and the quotient M/Γ is isomorphic to

$$X - \bigcup \{ S_j \mid b_j = \infty \}.$$

Remark: A uniformization is the universal uniformization if and only if it is simply connected.

Definition: Let X be an orbifold and M be its universal uniformization. The multivalued inverse map $X \to M$ of the projection $M \to X$ is called the <u>developing map</u>. It is uniquely determined up to the group $\text{Aut}(M)$ of holomorphic automorphisms of M.

Definition: A uniformization is called <u>finite</u> (respectively <u>abelian</u>) if the corresponding group K is finite (respectively if H/K is abelian).

§ 5.2 Uniformizing Differential Equations

Let us turn to the case where $X = \mathbb{P}_1$, $S = \{x_1, \ldots, x_m, x_{m+1} = \infty\}$, $b_j \in \{2, 3, \ldots, \infty\}$, and $\underline{X} = (X, S, b)$. If $b_j < \infty$, the local uniformization is given by $t \mapsto t^{b_j}$ and the corresponding

Galois group is $\mathbb{Z}/b_j\mathbb{Z}$. If $b_j = \infty$, then the local uniformization is given by the exponential function and the group is $\mathbb{Z}$. It is known (cf. [Kat 2]) that every orbifold (X,S,b) is uniformizable unless $m = 0$; or $m = 1$ and $b_1 \neq b_2$. In the case $m = 1$ and $b_1 = b_2 = b$ the developing map is globally given by

$$x \mapsto x^{1/b} \quad (b \neq \infty) \quad \text{or} \quad x \mapsto x^{1/\infty} = \log x \quad (b = \infty).$$

Now assume m is bigger than 1. Then X is uniformizable. Moreover in this case there is a finite uniformization whose existence was first proved by Fox([Fox 1]) who constructed it explicitly (the solution to <u>the Fenchel conjecture</u>). Later Selberg proved the following useful theorem of which Fox's theorem is a consequence.

<u>Theorem</u> ([Sel]): Any finitely generated subgroup Γ of $GL(n,\mathbb{C})$ has a normal subgroup Γ' of finite index (i.e. $\#(\Gamma/\Gamma') < \infty$) which is torsion free (i.e. there is no element of Γ' of finite order).

We here skip both the proofs of Selberg's theorem and of its corollary Fox' theorem. Now let M be the universal uniformization of $\underline{X}$. Since there are only three simply connected Riemann surfaces namely $\mathbb{P}_1$, $\mathbb{C}$, and the Poincaré upper half plane H, our universal uniformization M must be one of them. Since we have

$$\text{Aut}(\mathbb{P}_1) = PGL(2,\mathbb{C})$$

$$\text{Aut}(\mathbb{C}) = \left\{ \begin{pmatrix} a & b \\ 0 & 1 \end{pmatrix} \Big| \begin{array}{l} a \in GL(1,\mathbb{C}) \\ b \in \mathbb{C} \end{array} \right\} \cong GL(1,\mathbb{C}) \ltimes \mathbb{C}$$

$$\text{Aut}(H) = PU(1,1) \cong PSL(2,\mathbb{R})$$

We know that

$$\text{Aut}(M) \subset PGL(2,\mathbb{C})$$

Note: When a group G acts on a group H, the <u>semi-direct product</u> G ⋉ H of the groups G and H is the group defined as follows. As a set, it is G × H; and the product is defined by $(g,h) \cdot (g',h') = (g \cdot g', h \cdot g(h'))$.

Let x and z be coordinates of X and M, respectively. The developing map $z = z(x)$ is PGL(2,$\mathbb{C}$)-multivalued. Thus the Schwarzian derivative of the developing map is a single valued function on X. Set

$$q(x) = - \{z(x); x\}.$$

<u>Definition</u>: The uniquely determined differential equation in canonical form

$$\frac{d^2 u}{dx^2} + q(x) u = 0$$

is called <u>the uniformizing differential equation</u> of the orbifold (X,S,b).

The projective solution of the equation is the developing map $z(x)$. By the local behaviour of the Schwarzian derivative, we see that this is a Fuchsian differential equation with the following Riemann scheme:

$$\begin{pmatrix} x_1 & \cdots & x_m & x_{m+1} = \infty \\ \frac{1+1/b_1}{2} & \cdots & \frac{1+1/b_m}{2} & \frac{-1+1/b_{m+1}}{2} \\ \frac{1-1/b_1}{2} & \cdots & \frac{1-1/b_m}{2} & \frac{-1-1/b_{m+1}}{2} \end{pmatrix}$$

<u>Remark</u>: By § 2.6 Remark 2, when $m = 2$, the uniformizing equation is obviously projectively equivalent to the HGDE. The correspondence between the parameters (b_1, b_2, b_3) and (a,b,c) will appear soon. If $m \geq 3$ the uniformizing equation has accessory parameters, so that the local behavior of the developing map does not determine the equation. For this reason although the uniformizing equation is known to be

uniquely determined by the orbifold $\underline{X}$, unfortunately up to now nobody has been able to solve what is sometimes called the Poincaré problem, i.e. to write this equation explicitly.

§ 5.3 The Gauss-Schwarz Theory

We study now the developing map in the case $m = 2$. Let $X = \mathbb{P}_1$, $S = \{0,1,\infty\}$ and $b_0, b_1, b_\infty \in \{2, 3, \ldots, \infty\}$. By what proceeds we know that we have to study the HGDE :

$$x(x - 1)u'' + \{c - (a + b + 1)x\}u' - abu = 0.$$

Its Riemann scheme is

$$\begin{pmatrix} x = 0 & x = 1 & x = \infty \\ 0 & 0 & a \\ 1 - c & c-a-b & b \end{pmatrix}$$

The differences of the two characteristic exponents are, up to signs, the invariants of the equation under projective changes. Thus comparing the above Riemann scheme with that in canonical form, we get the equalities:

$$(1 - c)^2 = (\frac{1}{b_0})^2$$

$$(c - a - b)^2 = (\frac{1}{b_1})^2$$

$$(a - b)^2 = (\frac{1}{b_\infty})^2$$

Since the b_j's are integers (or ∞), a, b and c must be real numbers. We have

$$\frac{1}{b_0} = |1 - c|, \quad \frac{1}{b_1} = |c - a - b|, \quad \frac{1}{b_\infty} = |a - b|.$$

In order to understand the developing map $z = z(x) : X \to P_1$, we first restrict to the upper half plane $\{x \in \mathbb{C} | \operatorname{Im} x > 0\}$ of X. On account of Remark (4) in § 2.5, in an open neighborhood of the segment (0,1) such that its intersection with the real line is (0,1), there exist two linearly independent solutions which are real valued when restricted to (0,1). Thus around each of the intervals $(\infty,0)$, (0,1) and $(1,\infty)$, there is a projective solution real valued on the interval. Since circles or lines are mapped to circles or lines by projective transformations, and since projective solutions are projectively related, the image of the upper half plane by the projective solution is a curvilinear triangle bounded by circular arcs (<u>Schwarz's triangle</u>). Using the Frobenius method we can determine the angles of the triangle around $x = 0$ we have the following solutions

$$u_0 = 1 + \cdots = F(a,b,c;x)$$

and $\quad u_1 = \begin{cases} x^{1-c}(1 + \cdots) & \text{(when } c \neq 1) \\ u_0(\log x) + \text{holomorphic function} & \text{(when } c = 1). \end{cases}$

Thus the angle α_0 at $z = z(0)$ of the triangle is $\pi|1 - c|$. Similarly the angles α_1 and α_∞ at $z = z(1)$ and $z(\infty)$ are $\pi|c - a - b|$ and $\pi|a - b|$, respectively. Considering the circles C_∞, C_0 and C_1 producing the sides $(z(0),z(1))$, $(z(1),z(\infty))$ and $(z(\infty),z(0))$, respectively, we get the following picture of Schwarz's triangle.

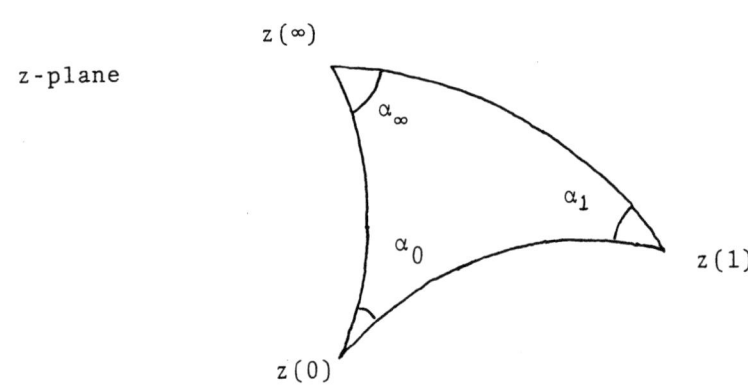

By Schwarz's reflection principle, the map $z = z(x)$ from the upper half plane to the triangle can be extended to the lower half plane through for example $(0,1)$. Then the image is the conformal reflection of the triangle with respect to the mirror C_∞. By repeating the same trick at any side of the already obtained polygon, we can extend $z = z(x)$ along any possible path in $X - S$. Let $\tilde{X}_z$ be the Riemann surface over X defined by the multivalued function $z(x)$. Then $x \to z(x)$ gives an isomorphism of $\tilde{X}_z$ with the multivalued image M under $z = z(x)$ of X. We have to distinguish the following three cases :

(i) $\quad \frac{1}{b_0} + \frac{1}{b_1} + \frac{1}{b_\infty} > 1 \quad$ then M is $\mathbf{P}_1$ itself.

(ii) $\quad \frac{1}{b_0} + \frac{1}{b_1} + \frac{1}{b_\infty} = 1 \quad$ then the three circles C_0, C_1 and C_∞ intersect at one point, say ∞, and $M = \mathbf{P}_1 - \{\infty\} \cong \mathbf{C}$.

(iii) $\quad \frac{1}{b_0} + \frac{1}{b_1} + \frac{1}{b_\infty} < 1 \quad$ then there exists a circle C orthogonal to the three circles and M is one of the open discs in $\mathbf{P}_1$ bounded by C.

In case (i) the projective monodromy group is a spherical triangle group, a finite subgroup of $PGL(2,\mathbf{C})$. There are four possibilities :

b_0	b_1	b_∞	
2	2	n	Dihedral group ($n = 2,3,\ldots$)
2	3	3	Tetrahedral group
2	3	4	Octahedral group
2	3	5	Icosahedral group

For the definition of these groups and the details of finite subgroups of $PGL(2,\mathbf{C})$, see § 11.2.

In case (ii) the projective monodromy group is a triangle group in $GL(1,\mathbb{C}) \ltimes \mathbb{C}$. There are four possibilities :

b_0	b_1	b_∞
2	2	∞
2	3	6
2	4	4
3	3	3

In case (iii) the monodromy group is a triange subgroup of $PU(1,1) \cong PSL(2,\mathbb{R})$. It is sometimes called a Schwarz triangle group. There are infinitely many possibilities. For instance the following groups are familiar :

b_0	b_1	b_∞	The projective monodromy group
2	3	∞	The modular group $\Gamma = PSL(2,\mathbb{Z})$
∞	∞	∞	The principal congruent subgroup of Γ of level 2.
2	3	7	The discrete subgroup of $PU(1,1)$ which has the smallest co-volume.

These three groups are arithmetic groups. There are finitely many arithmetic triangle subgroups which are tabulated in [Tu].

Here we show Schwarz's triangles for various cases.

< 2,2,n >

< 2,3,3 >

< 2,3,4 >

< 2,3,5 >

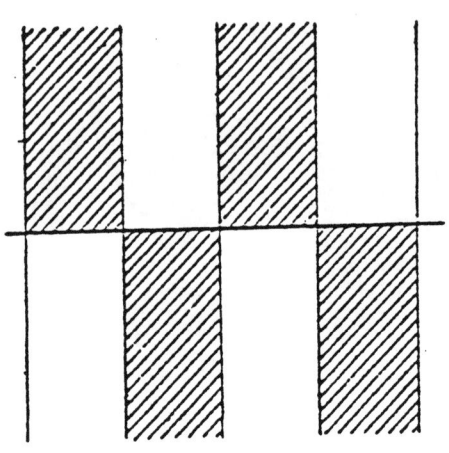

< 2,2,∞ >

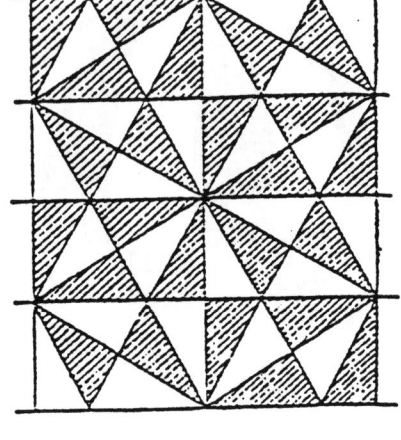

< 2,3,6 >

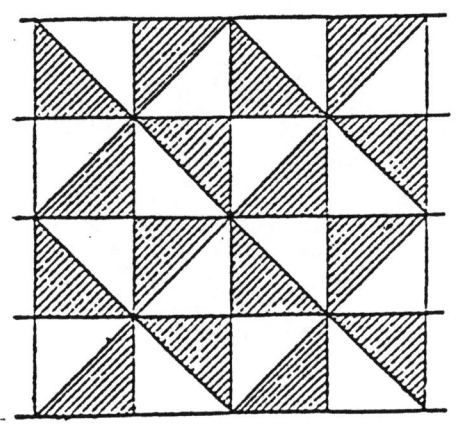

< 2,4,4 >

< 3,3,3 >

< 2, 3, ∞ >

< ∞, ∞, ∞ >

< 2, 3, 7 >

Part II

Chapter 6 Hypergeometric Differential Equations in Several Variables

This chapter introduces generalizations of the HGS and of the HGDE to N variables concentrating on the case N = 2. Appell's HGDE's are studied in detail.

§ 6.1 Hypergeometric Series

Following P.Appell, let us define the following four kinds of hypergeometric series in two variables:

$$F_1(a,b,b';c;x,y) = \Sigma \frac{(a,m+n)(b,m)(b',n)}{(c,m+n)(1,m)(1,n)} x^m y^n$$

$$F_2(a,b,b';c,c';x,y) = \Sigma \frac{(a,m+n)(b,m)(b',n)}{(c,m)(c',n)(1,m)(1,n)} x^m y^n$$

$$F_3(a,a',b,b';c;x,y) = \Sigma \frac{(a,m)(a',n)(b,m)(b',n)}{(c,m+n)(1,m)(1,n)} x^m y^n$$

$$F_4(a,b;c,c';x,y) = \Sigma \frac{(a,m+n)(b,m+n)}{(c,m)(c',n)(1,m)(1,n)} x^m y^n$$

where n and m run from 0 to infinity; and a,a',b,b',c and c' are complex numbers ($c,c' \neq 0,-1,-2,\ldots$). Continuing in this direction let us also define with Lauricella the HGS's in N variables:

$$F_D(a,b_1,..,b_N;c;x_1,..,x_N) = \Sigma \frac{(a,\Sigma n_j)\Pi(b_j,n_j)}{(c,\Sigma n_j)\Pi(1,n_j)} \Pi x_j^{n_j}$$

$$F_A(a,b_1,..,b_N;c_1,..,c_N;x_1,..,x_N) = \Sigma \frac{(a,\Sigma n_j)\Pi(b_j,n_j)}{\Pi(c_j,n_j)\Pi(1,n_j)} \Pi x_j^{n_j}$$

$$F_B(a_1,..,a_N,b_1,..,b_N;c;x_1,..,x_N) = \Sigma \frac{\Pi(a_j,n_j)\Pi(b_j,n_j)}{(c,\Sigma n_j)\Pi(1,n_j)} \Pi x_j^{n_j}$$

$$F_C(a,b;c_1,..,c_N;x_1,..,x_N) = \Sigma \frac{(a,\Sigma n_j)(b,\Sigma n_j)}{\Pi(c_j,n_j)\Pi(1,n_j)} \Pi x_j^{n_j}$$

Notice that when $N = 2$, the series F_D, F_A, F_B and F_C reduce to F_1, F_2, F_3 and F_4, respectively and that when $N = 1$ all of them reduce to $F(a,b,c;x)$. The series F_1 and F_3 converge in $\{|x| < 1, |y| < 1\}$ while F_2 converges in $\{|x| + |y| < 1\}$ and F_4 in $\{\sqrt{|x|} + \sqrt{|y|} < 1\}$.

Let us denote by $A_{m,n}$ the coefficient of $x^m y^n$ in any of the F_j's $(1 \leq j \leq 4)$ and put

$$f(m,n) = A_{m+1,n}/A_{m,n} \quad \text{and} \quad g(m,n) = A_{m,n+1}/A_{m,n}$$

Then we can write

$$f(m,n) = F(m,n)/F'(m,n) \quad \text{and} \quad g(m,n) = G(m,n)/G'(m,n)$$

where F and F' as well as G and G' are coprime polynomials in m and n. If we consider for instance the series F_1 then we have

$$F(m,n) = (a+m+n)(b+m) \quad , \quad G(m,n) = (a+m+n)(b'+n)$$
$$F'(m,n) = (c+m+n)(1+m) \quad , \quad G'(m,n) = (c+m+n)(1+n)$$

Furthermore f and g satisfy the compatibility condition :

$$f(m,n+1)g(m,n) = g(m+1,n)f(m,n)$$

It is known ([Sat]) that any rational functions f and g satisfiying the compatibility condition can be expressed by the ratios of products of linear forms in m and n which means for example that $F = \Pi_i(a_i m + b_i n + c_i)$. In particular assuming that F, F', G and G' are of degree two and that F'(m,n) and G'(m,n) have factors 1 + m and 1 + n, respectively, Horn(1931) obtained all the corresponding series.

$$F_j (j = 1,\ldots,4) , \quad G_j (j = 1,2,3) , \quad H_j (j = 1,\ldots,7),$$

which can be found in [Erd]. All these series are called the hypergeometric series (of degree two).

§ 6.2 Hypergeometric Differential Equations

Each hypergeometric series F (of degree two) satisfies a system of linear partial differential equations. In order to obtain these equations put

$$D = x\, \partial/\partial x \quad \text{and} \quad D' = y\, \partial/\partial y$$

and by using the arguments in § 1.2 check that the both operators

$$P_1 = P_1(x,y,D,D') = F(D,D') - F'(D,D')x^{-1}$$
and
$$P_2 = P_2(x,y,D,D') = G(D,D') - G'(D,D')y^{-1}$$

annihilate F. The following is the list of the P_1's and the P_2's for Appell's HGDE.

$F_1(a,b,b';c)$ $P_1 = x^{-1}D(c-1+D+D') - (a+D+D')(b+D)$

$P_2 = y^{-1}D'(c-1+D+D') - (a+D+D')(b'+D)$

$F_2(a,b,b';c,c')$ $P_1 = x^{-1}D(c-1+D) - (a+D+D')(b+D)$

$P_2 = y^{-1}D'(c'-1+D') - (a+D+D')(b'+D')$

$F_3(a,a',b,b';c)$ $P_1 = x^{-1}D(c-1+D+D') - (a+D)(b+D)$

$P_2 = y^{-1}D'(c-1+D+D') - (a'+D')(b'+D')$

$F_4(a,b;c,c')$ $P_1 = x^{-1}D(c-1+D) - (a+D+D')(b+D+D')$

$P_2 = y^{-1}D'(c'-1+D') - (a+D+D')(b+D+D')$

The system of partial differential equations

$$P_1 u = 0, \quad P_2 u = 0$$

is called the <u>hypergeometric differential equation</u> (HGDE) of degree two in two variables. Every system has the form

$$P_1 u = a^{11} u_{11} + a^{12} u_{12} + a^1 u_1 + a^2 u_2 + au = 0$$
$$P_2 u = b^{22} u_{22} + b^{21} u_{21} + b^2 u_2 + b^1 u_1 + bu = 0$$

where u_1 and u_2 denote the partial derivatives of u by x and y, respectively and u_{11} the second derivative with respect to x, and so on. By differentiating these equations repeatedly we can easily see that all the higher derivatives of u are expressed as linear combinations of u, u_1, u_2 and u_{12}. This shows that the dimension of the solution space of the system is at most 4. If the coefficients satisfy

(*) $\quad a^{11}b^{22} - a^{12}b^{21} = 0$

then the dimension of the solution space is at most 3. Indeed the equation

$$b^{22} \partial(P_1 u)/\partial y - a^{12} \partial(P_2 u)/\partial x = 0$$

gives rise to a linear combination of u, u_1, u_2 and u_{12}. For example, in the case of F_1, the two equations

$$P_1 u = x(1-x)u_{11} + y(1-x)u_{12} + (c-(a+b+1)x)u_1 - byu_2 - abu = 0$$

and

$$P_2 u = y(1-y)u_{22} + x(1-y)u_{21} + (c-(a+b'+1)y)u_2 - b'xu_1 - ab'u = 0$$

lead to

$$(x-y)u_{12} - b'u_1 + bu_2 = 0.$$

Actually every HGDE (of degree two) has a maximal solution space : of dimension 3 if (*) holds and of dimension 4 otherwise. For example, since F_j satisfies (*) only when j = 1, the solution spaces of Appell's equations F_1, F_2, F_3 and F_4 have dimension 3,4,4 and 4, respectively. This result is proved by using explicit power series in order to construct linearly independent local solutions (see [A-K]) , or by testing the integrability conditions of § 8.4 (see also [S-Y]). or also by using the Euler integrals presented in § 6.4 in order to construct global solutions.

Remark: Appell's HGDE's F_2 and F_3 are essentially the same in the following sense: If we transform the equation $F_3(a,a',b,b';c)$ (with the unknown z and the independent variable x and y) by letting x = 1/u, y = 1/v and z =

$x^{-b}y^{-b'}w$ then the equation with the unknown w and the independent variables u and v coincides with the equation $F_2(b+b'+1-c,b,b';b+1-a,b'+1-a')$

§ 6.3 Contiguity relations

As in the single variable case, if we increase (or decrease) any parameter of a HGS by one then it becomes a linear combination of the original HGS and of its derivatives. For Lauricella's F_D we need only the HGS and its first derivatives and have the following

<u>Proposition</u> (Contiguity relations ([Mil])): The following relations hold (as functions in $x_1,\ldots,x_N,a,b_1,\ldots,b_N$ and c)

$F_D(a+1,b_1,\ldots,b_N;c;x_1,\ldots,x_N)$
$= \frac{1}{a} (\Sigma\, x_i \frac{\partial}{\partial x_i} + a)\, F$

$F_D(a-1,b_1,\ldots,b_N;c;x_1,\ldots,x_N)$
$= \frac{1}{c-a} \{ \Sigma\, x_i(1-x_i) \frac{\partial}{\partial x_i} - \Sigma\, b_i x_i + c - a\}\, F$

$F_D(a,b_1,\ldots,b_{k-1},b_k+1,b_{k+1},\ldots,b_N;c;x_1,\ldots,x_N)$
$= \frac{1}{b_k} (x_k \frac{\partial}{\partial x_k} + b_k)\, F$

$F_D(a,b_1,\ldots,b_{k-1},b_k-1,b_{k+1},\ldots,b_N;c;x_1,\ldots,x_N)$
$= \frac{1}{c-\Sigma b_i} \{ x_k \Sigma\, (1-x_i) \frac{\partial}{\partial x_i} - a x_k + c - \Sigma b_i\}\, F$

$F_D(a,b_1,\ldots,b_N;c+1;x_1,\ldots,x_N)$
$= \frac{c}{(c-a)(c-\Sigma b_i)} \{ \Sigma\, (1-x_i) \frac{\partial}{\partial x_i} + c - a - \Sigma b_i\}\, F$

$F_D(a,b_1,\ldots,b_N;c-1;x_1,\ldots,x_N)$
$= \frac{1}{c-1} (\Sigma\, x_i \frac{\partial}{\partial x_i} + c - 1)\, F$

where $F = F_D(a,b_1,\ldots,b_N;c;x_1,\ldots,x_N)$.

Proof: If we write $b = (b_1,\ldots,b_N)$, $m = (m_1,\ldots,m_N)$,

$x^m = x_1^{m_1}\ldots x_N^{m_N}$, $A_m = \dfrac{(a,\Sigma m_i)\,\Pi(b_i,m_i)}{(c,\Sigma m_i)\,\Pi(1,m_i)}$ and $F = F_D(a,b;c;x)$

$= \Sigma\, A_m x^m$ then we have

$$F_D(a+1,b;c;x) = \Sigma\, A_m\, \frac{a+\Sigma m_i}{a}\, x^m$$

$$= \frac{1}{a}\,(\Sigma\, x_i\, \frac{\partial}{\partial x_i} + a)\, F.$$

The formulae for $F_D(a,b_1,\ldots,b_k+1,\ldots,b_N;c;x)$ and for $F_D(a,b;c-1;x)$ are obtained in the same way. In order to obtain the remaining relations, we use the following:

$$\{(1-x_j)\,\frac{\partial}{\partial x_j} - b_j\}\, F = (a-c)\, \Sigma_m\, A_m\, \frac{b_j+m_j}{c+\Sigma m_p}\, x^m$$

$$\{\Sigma_j(1-x_j)\,\frac{\partial}{\partial x_j} - a\}\, F = (\Sigma_j b_j - c)\, \Sigma_m\, A_m\, \frac{a+\Sigma m_p}{c+\Sigma m_p}\, x^m$$

which can be proved just as in § 1.3. We are now able to compute the last contiguity relations as follows:

$$F_D(a-1,b;c;x) = \Sigma\, A_m\, \frac{a-1}{a+\Sigma m_p - 1}\, x^m$$

$$= F - \Sigma\, A_m\, \frac{\Sigma m_p}{a+\Sigma m_p - 1}\, x^m$$

$$= F - \Sigma_m\, \Sigma_j\, A_{m_1,\ldots,m_j-1,\ldots,m_N}\, \frac{b_j+m_j-1}{c+\Sigma m_p - 1}\, x^m$$

$$= F - \Sigma_j\, x_j\, \Sigma_m\, A_m\, \frac{b_j+m_j}{c+\Sigma m_p}\, x^m$$

$$= F - \frac{1}{a-c}\, \Sigma_j\, x_j\{(1-x_j)\,\frac{\partial}{\partial x_j} - b_j\}\, F,$$

$$F_D(a,b;c+1;x) = \Sigma\, A_m \frac{c}{c+\Sigma m_p} x^m$$

$$= \Sigma\, A_m \frac{c}{\Sigma b_p - c} \left(\frac{\Sigma b_p + \Sigma m_p}{c + \Sigma m_p} - 1 \right) x^m$$

$$= \frac{c}{\Sigma b_p - c} \left\{ \frac{1}{a-c} \left(\Sigma(1-x_j) \frac{\partial}{\partial x_j} - \Sigma b_j \right) - 1 \right\} F,$$

$$F_D(a,b_1,\ldots,b_k-1,\ldots,b_N;c;x) = \Sigma\, A_m \frac{b_k - 1}{b_k + m_k - 1} x^m$$

$$= F - \Sigma\, A_m \frac{m_k}{b_k + m_k - 1} x^m$$

$$= F - \Sigma\, A_{m_1,\ldots,m_k-1,\ldots,m_N} \frac{a + \Sigma m_p - 1}{c + \Sigma m_p - 1} x^m$$

$$= F - \Sigma\, A_m x_k \frac{a + \Sigma m_p}{c + \Sigma m_p} x^m$$

$$= F + \frac{1}{c - \Sigma b_j} x_k \left\{ \Sigma(1-x_j) \frac{\partial}{\partial x_j} - a \right\} F$$

Thus the proof of the proposition is now completed.

§ 6.4 Euler's Integral Representations

The use of Euler's integrals that we are now defining give solutions of some of the HGDE's of degree two: $P_1 u = 0$, $P_2 u = 0$ (compare with § 1.4). For any complex numbers λ and λ' consider the following two kinds of Euler kernels

$$K(\lambda) = K(\lambda;s,t,x,y)$$
$$= (1-sx-ty)^{-\lambda} \qquad \text{(kernel of the first kind)}$$

and

$$K(\lambda,\lambda') = K(\lambda,\lambda';s,t,x,y)$$
$$= (1-sx)^{-\lambda}(1-ty)^{-\lambda'} \qquad \text{(kernel of the second kind)}$$

The Euler transformation $u(x,y)$ of a function $w(s,t)$ is given by the following formal integral

(1) $\quad u = \iint w(s,t) K(s,t,x,y) ds dt$

where $K = K(s,t,x,y)$ denotes the Euler kernel of the first or second kind. By "formal" we mean that no area of integration has been yet specified and that no convergence has been discussed. We want to find reasonably simple functions $w_0(s,t)$ whose Euler transformations are solutions of the HGDE's. In order to obtain these we need to find two operators $Q_j = Q_j(s,t,\theta,\theta')$ ($j = 1,2$) of first order, where $\theta = s\,\partial/\partial s$ and $\theta' = t\,\partial/\partial t$, and two functions $L_j = L_j(s,t,x,y)$ ($j = 1,2$) such that

$$P_j K = Q_j L_j \qquad (j = 1,2)$$

Indeed with such Q_j's and such L_j's we will have, modulo Note 2 below,

$$P_j u = \iint w(s,t)\, P_j K(s,t,x,y) ds dt$$
(2) $\quad\quad\quad = \iint w(s,t)\, Q_j L_j\, ds dt$
$$= \iint (Q_j^* w(s,t)) L_j\, ds dt + R_j$$

where the Q_j^*'s are the formal adjoint operators of the Q_j's and the R_j's the remainder terms defined in Note 1. By definition, the formal adjoint operator Q^* of any linear operator

$$Q = \Sigma\, a_{ij}(s,t)\, \frac{\partial^i}{\partial s^i}\, \frac{\partial^j}{\partial t^j}$$

is

$$Q^* = \Sigma\, (-1)^{i+j}\, \frac{\partial^i}{\partial s^i}\, \frac{\partial^j}{\partial t^j}\, a_{ij}(s,t).$$

Note: 1) By using integration by parts and Stokes' theorem several times, one sees that for any smooth functions f and g the following is true

$$\iint_D f\, Qg\, dsdt = \iint_D (Q^*f)\, g\, dsdt + R.$$

Here R is a remainder term obtained by integrating along the boundary ∂D of the area of integration D an expression depending on the restrictions to ∂D of f, of g, of the coefficients of Q and of some of their derivatives.

If the area can be chosen so that $R_j = 0$ (j = 1,2), the desired function $w_0(s,t)$ will be any of the solutions of the system of equations $Q_j^* w = 0$ (j = 1,2).

Note: 2) The reader must be aware that the first equality sign in the previous computation (2) is formal since one needs uniform convergence in (x,y) on the area of integration in order that (1) be defined and that differentiation in the (x,y)-direction under the integral sign be permitted.

Construction of Q_j's and L_j's (j = 1,2): Notice that for all complex numbers λ and λ' we have the following:

$$DK(\lambda) = \theta K(\lambda) \qquad D'K(\lambda) = \theta' K(\lambda)$$
$$(a+D+D')K(\lambda) = \lambda K(\lambda+1) + (a-\lambda)K(\lambda)$$
$$(1+D)x^{-1}K(\lambda) = \lambda sK(\lambda+1) \qquad (1+D')y^{-1}K(\lambda) = \lambda tK(\lambda+1)$$

$$DK(\lambda,\lambda') = \theta K(\lambda,\lambda') \qquad D'K(\lambda,\lambda') = \theta' K(\lambda,\lambda')$$
$$(a+D)K(\lambda,\lambda') = \lambda K(\lambda+1,\lambda') + (a-\lambda)K(\lambda,\lambda')$$
$$(a'+D')K(\lambda,\lambda') = \lambda' K(\lambda,\lambda'+1) + (a'-\lambda')K(\lambda,\lambda')$$
$$(1+D)x^{-1}K(\lambda,\lambda') = \lambda sK(\lambda+1,\lambda')$$
$$(1+D')y^{-1}K(\lambda,\lambda') = \lambda' tK(\lambda,\lambda'+1)$$

By using these formulae, we can, for some HGDE's, find Euler kernels K, operators Q_j of the first order and functions L_j (j = 1,2) such that $P_j K = Q_j L_j$ (j = 1,2). This is done simply by looking at the forms of the numerators of the coefficients $A_{m,n}$ of the HGS's. We know two situations in which the procedure succeeds.

(i) If the numerator of $A_{m,n}$ has a factor of the form (a,m+n), i.e. if $F(D,D')$ and $G(D,D')$ share a common factor of the form $(a+D+D')$, we choose $K = K(a)$ and take $L_1 = L_2 = K(a+1)$.

(ii) If the numerator of $A_{m,n}$ has a factor of the form (a,m)(a',n), i.e. $F(D,D')$ has a factor of the form $(a+D)$ and $G(D,D')$ has a factor of the form $(a'+D')$, we choose $K = K(a,a')$ and take $L_1 = K(a+1,a')$ and $L_2 = K(a,a'+1)$.

For instance, let us have a look at Appell's equation F_1. Since the numerator of $A_{m,n}$ is the product of (a,m+n), (b,m) and (b',n), both methods apply.

Method (i): $K = K(a)$ and $L_1 = L_2 = K(a+1)$. Since we have

$$P_1 K(a) = a((b+D) - s(c+D+D'))K(a+1)$$
$$= a((b+\theta) - s(c+\theta+\theta'))K(a+1)$$

we put

$$Q_1 = b + \theta - s(c + \theta + \theta').$$

Similarly we put

$$Q_2 = b' + \theta' - t(c + \theta + \theta').$$

By using the formulae $\theta^* = -(1 + \theta)$ and $(\theta')^* = -(1 + \theta')$

the formal adjoints of Q_1 and Q_2 are easy to obtain; we get:

$$Q_1^* = (s-1)\theta + s\theta' - cs + 3s + b - 1$$

and $$Q_2^* = (t-1)\theta' + t\theta - ct + 3t + b' - 1.$$

The system $Q_1^* w = 0$, $Q_2^* w = 0$ has, up to a multiplicative constant, a unique solution

$$w = s^{b-1} t^{b'-1} (1-s-t)^{c-b-b'-1}.$$

This implies that the integral

$$(3) \quad u = \iint s^{b-1} t^{b'-1} (1-s-t)^{c-b-b'-1} (1-sx-ty)^{-a} \, ds \, dt$$

is a formal solution of the system F_1.

<u>Method</u> (ii): $K = K(b,b')$, $L_1 = K(b+1,b')$ and $L_2 = K(b,b'+1)$. Since we have

$$P_1 K(b,b') = b\{(a+D+D') - (c+D+D')s\} K(b+1,b')$$
and $$P_2 K(b,b') = b'\{(a+D+D') - (c+D+D')t\} K(b,b'+1),$$

we put

$$Q_1 = a + \theta + \theta' - s(c + \theta + \theta')$$
and $$Q_2 = a + \theta + \theta' - t(c + \theta + \theta').$$

The formal adjoints are

$$Q_1^* = (s-1)\theta + (s-1)\theta' + 3s - cs + a - 2$$
and $$Q_2^* = (t-1)\theta' + (t-1)\theta + 3t - ct + a - 2.$$

The system $Q_1^* w = 0$, $Q_2^* w = 0$ has only a trivial solution (identically zero solution) when s and t are considered as independent variables. If however we set $s = t$, it acquires the solution $w = t^{a-1}(1-t)^{c-a-1}$, which implies that the <u>line</u> integral

$$u = \int t^{a-1}(1-t)^{c-a-1}(1-tx)^{-b}(1-ty)^{-b'} dt$$

gives a formal solution of the system F_1.

Following the same idea, if the coefficients of the series satisfy the condition (i) or (ii), we get the formal solutions expressed by Euler integrals. In order to get analytic expressions for these solutions, let us now specify the area (or paths) on which we want to integrate.

For example if

(*)   Re b > 0,   Re b' > 0   and   Re (c-b-b') > 0

then the integral

$$(**) \quad \iint_D s^{b-1} t^{b'-1} (1-s-t)^{c-b-b'-1} (1-sx-ty)^{-a} \, ds \, dt$$

(where D is the triangular region bounded by $s = 0$, $t = 0$ and $1-s-t = 0$) converges uniformly in x and y. Moreover (**) is a solution of the HGDE F_1 as we now show. If Re b, Re b' and Re(c-b-b') are sufficiently big then one can differentiate under the integral sign in the (x,y)-direction and the remainder terms $R_j (j = 1,2)$ in the formal computation (see Note 1 above) turn out to be zero and so integral (**) is a solution of the HGDE F_1. Since it is meromorphic in a, b, b' and c, by using analytic continuation, we see that provided (**) remains convergent, it is still a solution when a, b, b' and c vary.

We are not going to characterize here all the areas (or paths) that lead to solutions. We only specify several particularly useful and well known ones in order to produce the following formulae for each Appell's equation.

$F_1(a,b,b';c;x,y)$ $\quad\quad\quad\quad\quad\quad\quad\quad\quad\quad\quad$ ($|x| < 1$, $|y| < 1$)

$$= C_1 \iint_{D_1} s^{b-1} t^{b'-1} (1-s-t)^{c-b-b'-1} (1-sx-ty)^{-a} \, ds\, dt$$

$\quad\quad\quad\quad\quad\quad\quad\quad$ (Re $b > 0$, Re $b' > 0$, Re $(c-b-b') > 0$)

$$= C_1' \int_0^1 t^{a-1} (1-t)^{c-a-1} (1-tx)^{-b} (1-ty)^{-b'} \, dt$$

$\quad\quad\quad\quad\quad\quad\quad\quad\quad\quad\quad\quad$ (Re $a > 0$, Re $(c-a) > 0$)

$F_2(a,b,b';c,c';x,y)$ $\quad\quad\quad\quad\quad\quad\quad\quad\quad\quad$ ($|x| + |y| < 1$)

$$= C_2 \iint_{D_2} s^{b-1} t^{b'-1} (1-s)^{c-b-1} (1-t)^{c'-b'-1} (1-sx-ty)^{-a} \, ds\, dt$$

$\quad\quad\quad\quad$ (Re $b > 0$, Re $b' > 0$, Re $(c-b) > 0$, Re $(c'-b') > 0$)

$$= C_2' \iint_{D_2'} s^{a-c'} t^{a-c} (s+t-st)^{c+c'-a-2} (1-sx)^{-b} (1-ty)^{-b'} \, ds\, dt$$

$\quad\quad\quad\quad$ (Re$(a-c'+1) > 0$, Re$(a-c+1) > 0$, Re$(c+c'-a-1) > 0$)

$F_3(a,a',b,b';c;x,y)$ $\quad\quad\quad\quad\quad\quad\quad\quad\quad$ ($|x| < 1$, $|y| < 1$)

$$= C_3' \iint_{D_3'} s^{b-1} t^{b'-1} (1-s-t)^{c-b-b'-1} (1-sx)^{-a} (1-ty)^{-a'} \, ds\, dt$$

$\quad\quad\quad\quad\quad\quad\quad\quad$ ($b > 0$, Re $b' > 0$, Re $(c-b-b') > 0$)

$F_4(a,b;c,c';x,y)$ $\quad\quad\quad\quad\quad\quad\quad\quad\quad\quad$ ($\sqrt{|x|} + \sqrt{|y|} < 1$)

$$= C_4 \iint_{D_4} s^{a-c'} t^{a-c} (s+t-st)^{c+c'-a-2} (1-sx-ty)^{-b} \, ds\, dt$$

$\quad\quad\quad\quad$ (Re$(a-c'+1) > 0$, Re$(a-c+1) > 0$, Re$(c+c'-a-1) > 0$)

where

$C_1 = \Gamma(c)/\Gamma(b)\Gamma(b')\Gamma(c-b-b')$ $\quad\quad C_1' = \Gamma(c)/\Gamma(a)\Gamma(c-a)$

$C_2 = \Gamma(c)\Gamma(c')/\Gamma(b)\Gamma(b')\Gamma(c-b)\Gamma(c'-b')$

$C_2' = C_4 = (-1)^{c-a} \Gamma(c)\Gamma(c')/\Gamma(a-c)\Gamma(1-a+c)\Gamma(c+c'-a-1)$

$C_3' = \Gamma(c)/\Gamma(b)\Gamma(b')\Gamma(c-b-b')$

$D_1 = D_3' = \{(s,t) \in R^2 \mid s, t, 1-s-t \geq 0\}$

$$D_2 = \{ (s,t) \in \mathbb{R}^2 \mid 0 \leq s, t \leq 1 \}$$

$$D_2' = D_4 = \{ (s,t) \in \mathbb{R}^2 \mid 0 \leq s \leq 1, \ t \leq 0, \ s+t-st \geq 0 \}$$

In order to check the previous equalities, use

$$K(\lambda) = \Sigma \ \frac{(\lambda, m+n)}{(1,m)(1,n)} \ s^m t^n x^m y^n$$

$$K(\lambda, \lambda') = \Sigma \ \frac{(\lambda, m)(\lambda', n)}{(1,m)(1,n)} \ s^m t^n x^m y^n$$

$$\iint_{D_4} s^{a-c'+m} \ t^{a-c+n} (s+t-st)^{c+c'-a-2} \ dsdt$$

$$= (-1)^{c+2c'-a} \ \frac{\pi \ \Gamma(c+c'-a-1)}{\sin \pi (a-c) \ \Gamma(c) \Gamma(c')} \ \frac{(a, m+n)}{(c,m)(c',n)}$$

and the Dirichlet formula:

$$\iint_{D_1} s^{p-1} t^{q-1} (1-s-t)^{r-1} \ dsdt = \frac{\Gamma(p)\Gamma(q)\Gamma(r)}{\Gamma(p+q+r)}$$

Of course one wishes to get rid of the restrictions on the exponents of the integrands. This can be achieved by considering 2-dimensional version of the double contour loops. Indeed Hattori and Kimura ([H-K]) explicitly constructed these while Aomoto([Aom 3]) characterized them abstractly. For instance the cartesian product of two double contour loops around 0 and 1 — one in the s-plane, the other in the t-plane — may be used in place of the square D_2 in the previous table.

Remarks: 1) The integral expression for F_3 and the first integral expression for F_1 differ only by their kernels. The same remark applies to the expresson for F_4 and the second expression for F_2.

2) For some other HGDE's, we can find reasonably simple Euler integral representations but this is not always the case.

From these integral representaions we get much important informations. For instance we can compute the monodromy groups for Appell's equations (see [Nak] and [Kan]). In the case of F_1 the explicit results are given in § 11.6. Here we discuss the geometric meaning of the Euler integrals in general and find the singular loci of the Appell's equations in terms of divisors. The reader is asked to show that these divisors are indeed the singular loci of the equations. (The notion of singular locus will be presented in full in Chapter 8).

Let E be a HGDE the solutions of which are expressed by

$$\int\int w(s,t) K(s,t,x,y) \, ds \, dt$$

and let $\{w_j(s,t)\}_j$ be polynomials such that

$$w(s,t) = \Pi_j \, w_j(s,t)^{\mu_j}$$

for some complex numbers μ_j's. The integrand $w(s,t)$ determines the configuration

$$A_0 = \{(s,t) \in \mathbb{C}^2 | \, \Pi_j \, w_j(s,t) = 0 \}$$

in the (s,t)-plane while the kernel $K = K(s,t,x,y)$ determines in the (s,t)-plane the set $L_0(x,y)$ which is the line

$$1 - sx - ty = 0$$

when K is of the first kind, and the union of the two lines

$$1 - sx = 0 \quad \text{and} \quad 1 - ty = 0$$

when K is of the second kind.

Euler integrals of the first kind: Compactify the (s,t)-plane and the (x,y)-plane in order to obtain the projective planes $P_2(s,t)$ and $P_2(x,y)$. Let

$$A = A_0 \cup \{\text{the line at infinity}\}$$

and $\quad L(x,y,z) = \{(s,t,u) \in P_2(s,t) \mid uz - sx - ty = 0\}$

where (s,t,u) and (x,y,z) are homogeneous coordinates of $P_2(s,t)$ and $P_2(x,y)$, respectively; then the projective line $L(x,y,1)$ is the completion of the affine line $L_0(x,y)$. Consider the arrangement $A \cup L(x,y,z)$ in $P_2(s,t)$. The subset S of $P_2(x,y)$ where the arrangement changes its intersection pattern (i.e. the set of points (x,y,z) such that the line $L(x,y,z)$ is tangent to A or passes through a singular point of A) turns out to be the singular locus of the extension of the equation E to $P_2(x,y)$. The way the Euler kernel of the first kind has been defined suggests interpreting $P_2(x,y)$ as the dual space to $P_2(s,t)$ namely as the set of lines in $P_2(s,t)$. In this context, S becomes the dual curve to the curve A (i.e. the set of lines tangent to A or passing through a singular point of A).

The following is the list of the equations and pictures for the A's for the $L(x,y,z)$'s and for the S's determined by the Euler integrals of the first kind of Appell's equations F_1, F_2 and F_4. The dashed lines represent the $L(x,y,z)$'s.

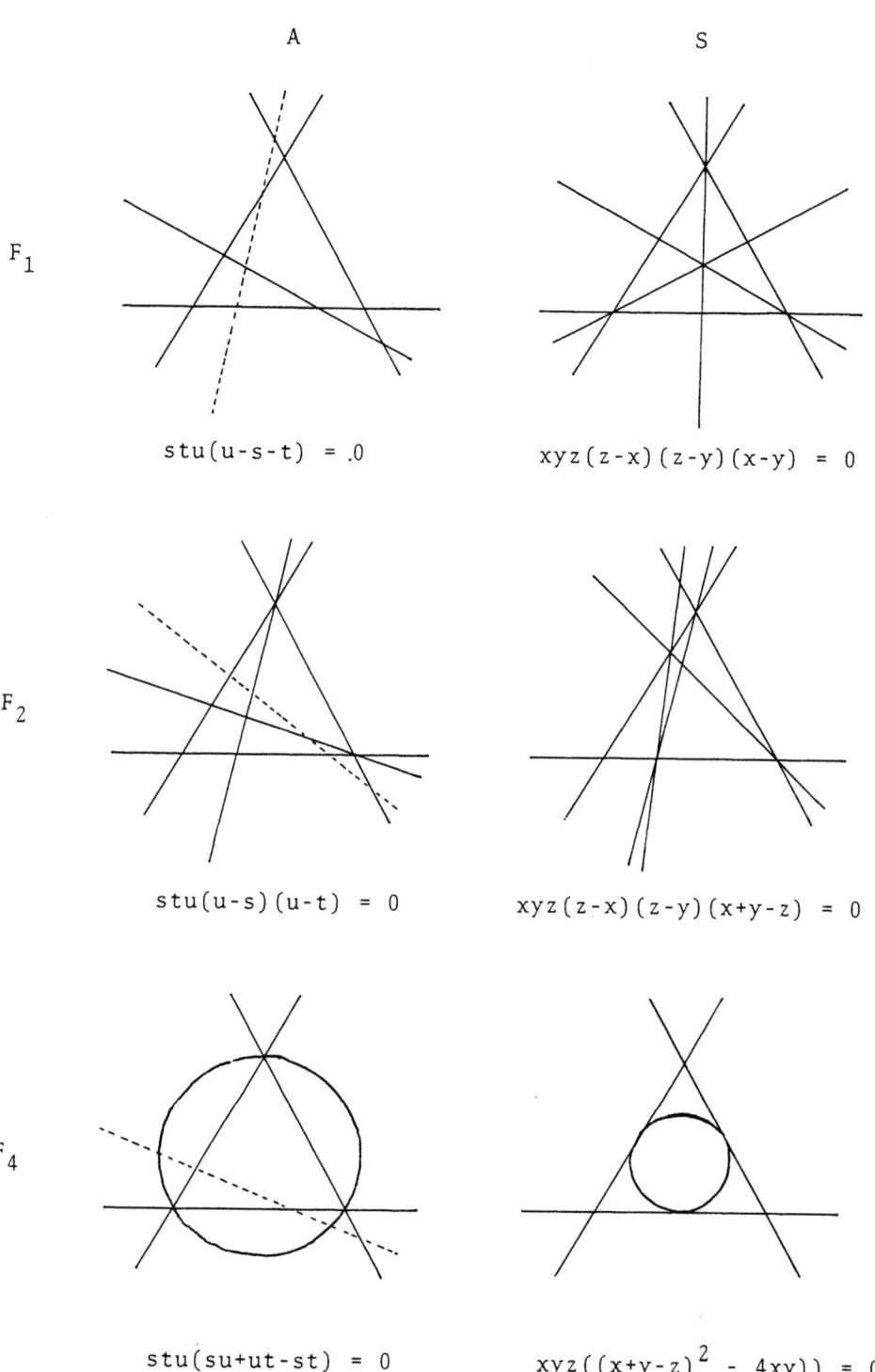

Euler integrals of the second kind: Compactify the (s,t)-plane and the (x,y)-plane in order to obtain the products of projective lines $P_1(s) \times P_1(t)$ and $P_1(x) \times P_1(y)$. Let

$$A = A_0 \cup \{x = \infty\} \cup \{y = \infty\}$$

and $\quad L(x,y) = \{(s,t) \in P_1(s) \times P_1(t) \mid s = x^{-1} \text{ or } t = y^{-1}\}$

for $(x,y) \in P_1(x) \times P_1(y)$. Notice that if x^{-1} and y^{-1} are finite, $L(x,y)$ is the completion of $L_0(x,y)$. Consider the arrangement $A \cup L(x,y)$ in $P_1(s) \times P_1(t)$. Let S be the subset of $P_1(x) \times P_1(y)$ where the arrangement changes its intersection pattern (or equivalently the set of (x,y) such that $L(x,y)$ and A have a contact of order greater of equal to two). In other words S is the set of (x,y) such that one of the two lines in $L(x,y)$ is tangent to A or passes through a singular point of A or such that both lines pass through the same point of A. Then S turns out to be the singular locus of the extension of the equation E to $P_1(x) \times P_1(y)$. The way that Euler kernel of the second kind has been defined suggests that one interprets $P_1(x) \times P_1(y)$ as the dual space to $P_1(s) \times P_1(t)$ namely as the set of pairs of lines $L(x,y)$ in $P_1(s) \times P_1(t)$. In this context, S is seen to be the dual curve to A (i.e. the set of pairs of lines $L(x,y)$ having contact at least two with A).

Here are the equations and pictures for the A's, for the A's, for the $L(x,y)$'s, and for the S's determined by the Euler integral representations of the second kind of Appell's equations F_2 and F_3. The dashed lines represent the $L(x,y)$'s. (The lines at infinity are omitted.)

	A	S
F_2	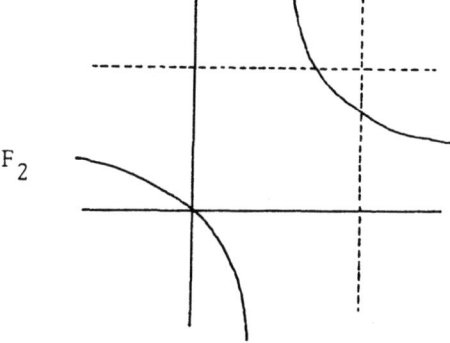 $st(s+t-st) = 0$	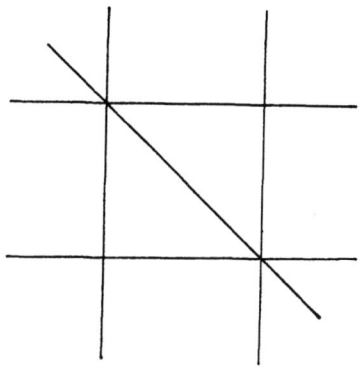 $xy(1-x)(1-y)(x+y-1) = 0$
F_3	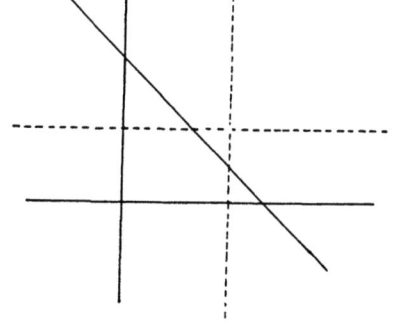 $st(1-s-t) = 0$	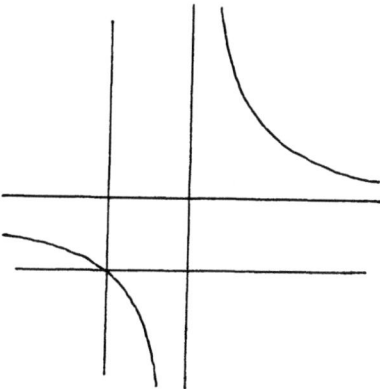 $xy(1-x)(1-y)(xy-x-y) = 0$

§ 6.5 The Erdelyi-Takano Integral Representation of F_C

Let F_C^N denote Lauricella's HGS in N variables. The following (line) integral representation of F_C^{N+1} involving F_C^N in its integrand is due to K. Takano.

$$F_C^{N+1}(a,b;c_1,\ldots,c_{N+1};x_1,\ldots,x_N)$$
$$= -4^{-1}\pi^{-2} \exp(\pi i(c_N+c_{N+1})) \Gamma(c_N)\Gamma(c_{N+1})\Gamma(2-c_N-c_{N+1})$$
$$\times \int_\gamma t^{-c_N}(1-t)^{-c_{N+1}}$$
$$\times F_C^N(a,b;c_1,\ldots,c_{N-1},c_N+c_{N+1}-1;x_1,\ldots,x_{N-1},\frac{x_N}{t}+\frac{x_{N+1}}{1-t})dt$$

where γ is the double contour loop around $t = 0$ and 1. It is a generalization of Erdelyi's integral representation of F_4 involving the HGF (in one variable). With the aid of this integral representation in the case $N = 2$, Takano computed the monodromy representation of F_4 (see [Tkn]) in which a method to compute the monodromy of F_C^N is also suggested.

§ 6.6 The Barnes Integral Representations

As well as in the single variable case the HGDE's can be transformed by using the Mellin transformation

$$u = \iint w(s,t) x^s y^t \, ds \, dt$$

into a system of difference equations, which can be solved by ratios of products of Gamma functions. For example we have

$$F_1(a,b,b';c;x,y) = -\frac{\Gamma(c)}{\Gamma(a)\Gamma(b)\Gamma(b')} \frac{1}{4\pi^2}$$

$$\times \iint_{-i\infty}^{+i\infty} \frac{\Gamma(a+s+t)\Gamma(b+s)\Gamma(b'+t)}{\Gamma(c+s+t)} \Gamma(-s)\Gamma(-t)(-x)^s(-y)^t \, dsdt$$

$$F_2(a,b,b';c,c';x,y) = -\frac{\Gamma(c)\Gamma(c')}{\Gamma(a)\Gamma(b)\Gamma(b')} \frac{1}{4\pi^2}$$

$$\times \iint_{-i\infty}^{+i\infty} \frac{\Gamma(a+s+t)\Gamma(b+s)\Gamma(b'+t)}{\Gamma(c+s)\Gamma(c'+t)} \Gamma(-s)\Gamma(-t)(-x)^s(-y)^t \, dsdt$$

$$F_3(a,a',b,b';c;x,y) = -\frac{\Gamma(c)}{\Gamma(a)\Gamma(a')\Gamma(b)\Gamma(b')} \frac{1}{4\pi^2}$$

$$\times \iint_{-i\infty}^{+i\infty} \frac{\Gamma(a+s)\Gamma(a'+s)\Gamma(b+s)\Gamma(b'+t)}{\Gamma(c+s+t)} \Gamma(-s)\Gamma(-t)(-x)^s(-y)^t dsdt$$

$$F_4(a,b;c,c';x,y) = -\frac{\Gamma(a)\Gamma(b)}{\Gamma(c)\Gamma(c')} \frac{1}{4\pi^2}$$

$$\times \iint_{-i\infty}^{+i\infty} \frac{\Gamma(a+s+t)\Gamma(b+s+t)}{\Gamma(c+s)\Gamma(c'+t)} \Gamma(-s)\Gamma(-t)(-x)^s(-y)^t dsdt$$

§ 6.7 A Relation Between the Equation F_D and the Garnier System G_N

There is an interesting relation between the HGF F_D and the Garnier system G_N which governs isomonodromic deformations (§ 3.6). In the theorem below, we assert without proof that under a given restriction on the parameters the Garnier system G_N admits solutions expressible in terms of the HGF F_D in N variables. In order to state the theorem we need the following

Lemma ([K-O]): The "canonical" transformation

$$s_j = \frac{t_j}{t_j - 1}, \quad q_j = t_j M_j, \quad p_j = -(t_j - 1) \sum_k \frac{M^{k,j} \mu_k}{\lambda_k(\lambda_k - 1)}$$

carries G_N to the Hamiltonian system

$$G_N' : \quad \frac{\partial q_i}{\partial s_j} = \frac{\partial h_j}{\partial p_i} \quad , \quad \frac{\partial p_i}{\partial s_j} = - \frac{\partial h_j}{\partial q_i} \qquad (1 \le i,j \le N)$$

whose Hamiltonians are

$$h_j = \frac{1}{s_j(s_j-1)} \{ \sum_{i,k} E^j_{ik}(s,q) p_i p_k - \sum_i F^j_i(s,q) p_i + \kappa q_j \}$$

($1 \le j \le N$), where the E's and the F's are polynomials in $q_1,\ldots,q_N$ of degree 3 and 2, respectively, such that $E^j_{ik} = E^j_{ki}$, $F^j_i = F^i_j$. One has

$$F^i_j = (\kappa_0 - 1) q_j (q_j - 1) + \kappa_1 q_j (q_j - s_j) + \theta_j (q_j - 1)(q_j - s_j)$$

$$+ \sum_{\substack{k=1 \\ k \ne j}}^{N} \{ \theta_k q_j (q_j - \frac{s_j(s_k-1)}{s_k - s_j}) - \theta_j \frac{s_j(s_j-1)}{s_j - s_k} q_k \} \qquad \text{if } i = j$$

$$F^i_j = (\kappa_0 + \kappa_1 + \sum_{k=1}^{N} \theta_k - 1) q_j q_i$$

$$+ \theta_i \frac{s_i(s_j-1)}{s_i - s_j} q_j + \theta_j \frac{s_j(s_i-1)}{s_j - s_i} q_i \qquad \text{if } i \ne j$$

where $\theta_j, \kappa_0, \kappa_1, \rho_\infty$ and κ_∞ are constants defined in § 3.6. In the case $\kappa = 0$ observe that when $p_j = 0$ ($1 \le j \le N$) in G_N' the second group of equations $\partial p_i / \partial s_j = - \partial h_j / \partial q_i$ becomes trivial while the first reduces to

$$(*) \quad \frac{\partial q_j}{\partial s_j} = \frac{1}{s_j(s_j-1)} F^j_i(s,q) \qquad (1 \le i,j \le N).$$

Since the $F^j_i(s,q)$'s are polynomials of degree 2 in the q_j's, by using a routine technique (the one which transforms

Riccati's equations into linear equations), the system (*) can be reduced to a system of linear equations, so that we have

<u>Theorem</u> ([O-K]): In the case $\kappa = 0$, the system G_N' admits solutions of the form

$$(q_1,\ldots,q_N,p_1,\ldots p_N) = (q_1(s),\ldots,q_N(s),0,\ldots,0)$$

$$q_j(s) = \frac{s_j(s_j-1)}{\kappa_0+\kappa_1+\Sigma\theta_m-1} \frac{\partial \log\{(s_j-1)^{\theta_j} u(s)\}}{\partial s_j}$$

where the auxiliary function $u(s)$ is an arbitrary solution of the equation $F_D(a,b_1,\ldots,b_N;c;s_1,\ldots,s_N)$ with $a = 1 - \kappa_1$, $b_j = \theta_j$ and $c = \kappa_0 + \Sigma\theta_m$.

§ 6.8 Confluent Hypergeometric Equations

As in the single variable case, by taking limits of HGDE's we are able to produce some confluent HGDE's. Several classical confluent HGDE's can be found in [Erd]. In § 3.6 and § 6.7 we saw that the isomonodromic deformations of Fuchsian equations of the second order are governed by the Hamiltonian system G_N and that when $\kappa = 0$ the system G_N can be solved by using the HGF F_D. Starting from confluent ordinary differential equations of the second order we obtain confluent versions of G_N and then confluent versions of Lauricella's F_D. In [O-K], K. Okamoto and H. Kimura tabulated explicit forms of equations as well as integral formulae for the confluent equations derived from F_1.

Chapter 7 The General Theory of Differential Equations

In this chapter we recall some basic facts about systems of linear partial differential equations in the single unknown on open subsets of $\mathbb{C}^n$. Since the ring of linear partial differential operators is not a principal ideal domain when $n \geq 2$, the situation is not so simple as in the single variable case. To develop the theory in full it is natural to discuss it at the sheaf level but in order to make this chapter elementary we avoid the introduction of unnecessarily abstract general notions. We use elementary algebro-geometric language to introduce the minimal amount of material which we need. We also discuss differential equations defined on complex manifolds. The general background to this chapter is in [SKK].

§ 7.1 Singularities of Differential Equations

Let $x = (x_1, \ldots, x_n)$ be a coordinate system on $\mathbb{C}^n$, let U be an open subset of $\mathbb{C}^n$ and let $O(U)$ be the ring of holomorphic functions on U. Let D_j ($1 \leq j \leq n$) be the differential operator $\partial/\partial x_j$ and let

$$D(U) = O(U)[D_1, \ldots, D_n]$$

be the ring of linear partial differential operators with coefficients in $O(U)$. To a given <u>ideal</u> I of $D(U)$ (by which we will always mean a left ideal) we associate the set $Iu = 0$ of differential equations $Pu = 0$ ($P \in I$) in the unknown u.

It is convenient to consider a $D(U)$-module $M(I) = D(U)/I$. Since the ring $D(U)$ is noetherian, the set $Iu = 0$ reduces to the system $P_j u = 0$ $(j = 1,..,t)$ where $\{P_1,...,P_t\}$ is a finite set of generators of I. Notice that if $n \geq 2$, in general there is no canonical way to choose generators. Thus the number t of generators and the orders of the operators P_j have no intrinsic meaning.

An operator $P \in D(U)$ of order m is expressed as

$$P = P(x,D) = \Sigma_{|k| \leq m} a_k(x) D^k$$

where

$$k = (k_1,..,k_n), \quad |k| = k_1 + ... + k_n$$

and $D^k = D_1^{k_1} ... D_n^{k_n}$.

Let T^*U be the cotangent bundle of U and let $\xi_1,..., \xi_n$ be a coordinate system on its fiber at $(x_1,..,x_n)$. The symbol $\sigma(P)$ of P is the function on T^*U defined by

$$\sigma(P) = \Sigma_{|k|=m} a_k(x) \xi_1^{k_1} ... \xi_n^{k_n}.$$

For every ideal I of $D(U)$ let the ideal of symbols of I be defined by putting $\sigma(I) = \{\sigma(P) | P \in I\}$; check that this set is indeed a homogeneous ideal of $O(U)[\xi_1,..., \xi_n]$. It should be noted that even if the operators $P_1,...,P_t$ generate the ideal I, the polynomials $\sigma(P_1),..., \sigma(P_t)$ do not necessarily generate the ideal $\sigma(I)$. The characteristic variety $Ch(I)$ of I is the subvariety of T^*U defined by the ideal $\sigma(I)$.

Let the singular locus of the equation $Iu = 0$ be the analytic subset of U defined by

$$S(I) = \pi(Ch(I) - U \times \{0\})$$

where $\pi : (x,\xi) \mapsto x$ is the projection $T^*U \to U$ of the cotangent bundle and $U \times \{0\}$ is the zero section.

Since in the present book we do not consider hyperfunction solutions, our set of solutions for a differential operator P is not modified if we multiply or divide P on the left by any non-zero element in O(U). For this reason, it is natural in the present context to extend a given ideal I of D(U) to the greatest ideal $\tilde{I}$ such that $\tilde{I}u = 0$ has the same solutions as $Iu = 0$. This is achieved by putting

$$\tilde{I} = \{P \in D(U) \mid aP \in I \text{ for some } a \in O(U) - \{0\}\}.$$

Now the O(U)-module $D(U)/\tilde{I}$ has no torsion and $S(\tilde{I})$ ($\subset S(I)$) is a subvariety of codimension 1 in U.

Let V be an another open subset of $\mathbb{C}^n$ contained in U. We define <u>the restriction</u> of an ideal I of D(U) to V by

$$I \mapsto I|_V = O(V) \underset{O(U)}{\otimes} I$$

which gives rise to an injective map from the set of ideals I of D(U) with $I = \tilde{I}$ into the set of ideals J of D(V) with $J = \tilde{J}$. For an ideal J of D(V), if there is an ideal I of D(U) whose restriction to V is J, the ideal I is called an <u>extention</u> of J. If U - V is contained in a subvariety of U, then the image of the restriction is the set of ideals of D(V) generated by operators which are extendable to U as differential operators with meromorphic coefficients. In this case the map

$$J \mapsto \{fP \in D(U) \mid f \in O(U), P \in J\}$$

gives an extention, which sends ideals J of D(V) with $J = \tilde{J}$ to ideals I of D(U) with $I = \tilde{I}$.

§ 7.2 Holonomic Systems

Let I be an ideal of $D(U)$, let $V = U - S(I)$ and let J be the restriction of I to V. Then the $D(V)$-module $M(V) = D(V)/J$ is a free $O(V)$-module the rank $r(I)$ of which is called the rank of the ideal I. Notice that any restriction and any extension of I has the same rank as I and that $r(I) = r(\tilde{I}) \in \mathbb{N} \cup \{\infty\}$. The ideal I as well as the equation $Iu = 0$ is said to be holonomic if $r(I)$ is finite. If the ideal is holonomic then there are $r(I)$-linearly independent holomorphic solutions around each point in $X - S(I)$. It is known (see [SKK]) that I is holonomic if and only if $\dim Ch(I) = n$.

Pfaffian Systems: Let I be a holonomic ideal of $D(U)$, $V = U - S(I)$, $J = I|_V$ and let $(\delta(i))$ ($1 \le i \le r(I)$) be a system of elements in $D(U)$ which after quotienting by I and restricting to V becomes a basis of the $O(V)$-free module $M(J) = O(V) \otimes_{O(U)} M(U) = O(V) \otimes_{O(U)} (D(U)/I)$. Since the $O(V)$-module $M(J) = D(V)/J$ is also a $D(U)$-module, $D_k \delta(i)$ is a $O(V)$-linear combination of the $\delta(j)$'s modulo J, in other words, there exist c_{kj}^i's in $O(V)$ such that

$$D_k \delta(i) \equiv \sum_{j=1}^{r(I)} c_{kj}^i \delta(j) \qquad \text{modulo } J.$$

This implies that if we introduce $r(I)$ unknowns $u^j = \delta(j)u$ ($1 \le j \le r(I)$) then the equation $Ju = 0$ is transformed into the system of first order linear partial differential equations

$$D_k u^i = \sum_{j=1}^{r(I)} c_{kj}^i u^j \qquad (1 \le k, i \le r(I))$$

or as specialists like to say into the $r(I)$-Pfaffian system

$$d\,{}^t(u^1,\ldots,u^{r(I)}) = \Omega\,{}^t(u^1,\ldots,u^{r(I)})$$

where $\Omega = (\Sigma c_{kj}^i dx^k)_{i,j=1}^{r(I)}$.

If $r(I) \geq 1$ we can take $\delta(1) = 1$, that is to say $u^1 = u$.

<u>Computing</u> $r(I)$: Unfortunately one knows no effective algorithm which would compute the rank of the ideal $I \subset D(U)$ generated by a given set of operators $P_1, \ldots, P_t$. But for some differential equations, the following procedure actually works. Suppose indeed that one has already been able to find $r-1$ $O(U)$- linear combinations $u_2, \ldots, u_r$ of the derivatives (of arbitrary order) of u_1 such that the system $Iu_1 = 0$ can be transformed into the r-Pfaffian system

(7.1) $d\,{}^t(u_1, \ldots, u_r) = \Omega\,{}^t(u_1, \ldots, u_r)$.

Suppose also that Ω is an $r \times r$ matrix whose entries ω_j^i ($1 \leq i, j \leq r$) are 1-forms on U the coefficients of which are rational functions in the coefficients of the P_j's and some of their derivatives and therefore belonging to the field of fractions of $O(U)$. Now all r-<u>Pfaffian systems</u> (i.e. systems of the form (7.1)) have at most r linearly independent vector valued solutions so that in our case we get

(7.2) $r \geq r(I)$.

Moreover (by Frobenius' theorem) the r-Pfaffian system (7.1) has exactly r linearly independent vector valued solutions if and only if it is <u>integrable</u> i.e. if and only if it satisfies the <u>integrability condition</u>

(7.3) $d\Omega - \Omega \wedge \Omega = 0$

(where $\Omega \wedge \Omega$ is the $r \times r$ matrix whose (i,j)-th entry is the 2-form $\Sigma_{k=1}^r \omega_k^i \wedge \omega_j^k$). In our case this means that (7.3) implies $r = r(I)$ and that $S(\tilde{I})$ ($\subset U$) is contained in the set of poles of the entries of the matrix Ω.

Examples: 1) By adding to the unknown $u = u_1$ of the HGDE F_J defined on $U = \mathbb{C}^n$ (where J stands for A, B or C) the $2^n - 1$ derivatives

$$D_{j_1}u \ (1 \leq j_1 \leq n), \ D_{j_1}D_{j_2}u \ (1 \leq j_1 < j_2 \leq n),$$
$$\ldots, D_1\ldots D_n u$$

one is led to an integrable 2^n-Pfaffian system so that in this case $r(I) = 2^n$.

2) Similarly, adding to the unknown $u = u_1$ of the HGDE F_D (for which again $U = \mathbb{C}^n$) the n derivatives $D_j u \ (1 \leq j \leq n)$ one can check that $r(I) = n+1$ (see § 7.3 and § 7.4).

Remark: Suppose that, by adding $O(U)$-linear combinations of partial derivatives of u, you have succeeded in finding a Pfaffian system (7.1) equivalent to the system $Iu = 0$ you are studying. If your (7.1) does not satisfy integrability condition (7.3), you need not give up. Indeed it is worth trying repeatedly to modify and eliminate some of the auxiliary functions you added provided that the reduced system remains of the form (7.1). With a certain amount of luck you might eventually end up with a reduced system (7.1) satisfying (7.3). This is the method we used to get the minimal number of partial derivatives needed in Examples 1) and 2).

§ 7.3 The Equation F_1

In this section, we apply the notions that have just been introduced to Appell's equation F_1. Since F_1 is defined on $U = \mathbb{C}^2$, we may write D and O in place of $D(U)$ and $O(U)$. Let I be the ideal of D generated by P_1 and P_2 where

$$P_1 = x_1(1-x_1)D_1^2 + x_2(1-x_1)D_1D_2 + (c-(a+b+1)x_1)D_1$$
$$- bx_2D_2 - ab$$

and
$$P_2 = x_2(1-x_2)D_2^2 + x_1(1-x_2)D_1D_2 + (c-(a+b'+1)x_2)D_2$$
$$- b'x_1D_1 - ab'$$

Check that $I = \tilde{I}$. The symbols of the P_j's are

$$\sigma(P_1) = x_1(1-x_1)\xi_1^2 + x_2(1-x_1)\xi_1\xi_2$$
$$= (x_1(1-x_1)\xi_1 + x_2(1-x_1)\xi_2)\xi_1$$

and
$$\sigma(P_2) = x_2(1-x_2)\xi_2^2 + x_1(1-x_2)\xi_1\xi_2$$
$$= (x_2(1-x_2)\xi_2 + x_1(1-x_2)\xi_1)\xi_2$$

Notice that the ideal $\sigma(I)$ of symbols of I is not generated by $\sigma(P_1)$ and $\sigma(P_2)$ alone; we need in addition a third generator $\sigma(P_3)$ where, say

$$P_3 = (1-x_2)D_2P_1 - (1-x_1)D_1P_2 + bP_2 - b'P_1$$
$$= (c-a-1)((x_1-x_2)D_1D_2 - b'D_1 + bD_2)$$

and hence $\sigma(P_3) = (c-a-1)(x_1-x_2)\xi_1\xi_2$.

Thus although it is of codimension 2, the characteristic variety $Ch(I)$ is the intersection of the three hypersurfaces $\sigma(P_j) = 0$ ($j = 1,2,3$). This variety decomposes into the six following irreducible components

$$Ch(I) = \{x_1-x_2 = \xi_1+\xi_2 = 0\} \cup \{x_2 = \xi_1 = 0\}$$
$$\cup \{1-x_2 = \xi_1 = 0\} \cup \{x_1 = \xi_2 = 0\}$$
$$\cup \{1-x_1 = \xi_2 = 0\} \cup \{\xi_1 = \xi_2 = 0\}$$

By projecting on U, one gets the singular locus :

$$S(I) = \{x_1(1-x_1)\ x_2(1-x_2)(x_1-x_2) = 0\}$$

Let $V = U - S(I)$ then the $O(V)$-module $O(V) \underset{O}{\otimes} M(I)$ admits the $O(V)$-free bases $1, D_1$ and D_2 so that the rank $r(I)$ is 3.

Transform the equation $Iu = 0$ by putting $u^1 = u$, $u^2 = x^1 D_1 u$, $u^3 = x^2 D_2 u$ in order to get the Pfaffian system

$$d\ {}^t(u^1, u^2, u^3) = \Omega\ {}^t(u^1, u^2, u^3)$$

where

$$\Omega = A_1 \frac{dx}{x} + A_2 \frac{dy}{y} + A_3 \frac{dx}{x-1} + A_4 \frac{dy}{y-1} + A_5 \frac{d(x-y)}{x-y}$$

and $x = x^1,\ y = x^2$

$$A_1 = \begin{bmatrix} 0 & 1 & 0 \\ 0 & 1-c+b' & 0 \\ 0 & -b' & 0 \end{bmatrix}, \quad A_2 = \begin{bmatrix} 0 & 0 & 1 \\ 0 & 0 & -b \\ 0 & 0 & 1-c+b \end{bmatrix}$$

$$A_3 = \begin{bmatrix} 0 & 0 & 0 \\ -ab & c-a-b-1 & -b \\ 0 & 0 & 0 \end{bmatrix}, \quad A_4 = \begin{bmatrix} 0 & 0 & 0 \\ 0 & 0 & 0 \\ -ab' & -b' & c-a-b'-1 \end{bmatrix}$$

$$A_5 = \begin{bmatrix} 0 & 0 & 0 \\ 0 & -b' & b \\ 0 & b' & -b \end{bmatrix}$$

Check that $d\Omega - \Omega \wedge \Omega = 0$.

§ 7.4 Equations of Rank $n+1$

Although in general there is no canonical way to choose generators of an ideal I of $D(U)$, when I is holonomic of rank $n+1$ it is convenient to generate it by n^2 elements P_{ij} of the form

$$P_{ij} = a_{ij} D_i D_j + \Sigma_{k=1}^n a_{ij}^k D_k + a_{ij}^0 \qquad i,j = 1,\ldots,n$$

where a_{ij}, a_{ij}^k, $a_{ij}^0 \in O(U)$. Conversely an ideal I generated by some family (P_{ij}) of operators of the previous form with $a_{ij} \in O(U) - \{0\}$ is holonomic of rank $\leq n+1$, since outside $S(I)$, the $O(U)$-module $M(U)$ is generated by $1, D_1, \ldots, D_n$. The ideal I is of rank $n+1$ if and only if

$$(7.4) \quad p_{ij}^k = p_{ji}^k \quad , \quad p_{ij}^0 = p_{ji}^0$$

$$D_k p_{ij}^m + \Sigma_{t=1}^n p_{ij}^t p_{kt}^m + \delta_k^m p_{ij}^0$$
$$= D_i p_{kj}^m + \Sigma_{t=1}^n p_{kj}^t p_{it}^m + \delta_i^m p_{kj}^0$$

$$D_k p_{ij}^0 + \Sigma_{t=1}^n p_{ij}^t p_{kt}^0 = D_i p_{kj}^0 + \Sigma_{t=1}^n p_{kj}^t p_{it}^0$$

$$(1 \leq i,j,k,m \leq n)$$

in which the p_{ij}^k's and the p_{ij}^0's are the coefficients of the differential operators

$$P'_{ij} = P_{ij}/a_{ij} = D_i D_j - \Sigma_{k=1}^n p_{ij}^k D_k - p_{ij}^0$$

i.e.

$$p_{ij}^k = -a_{ij}^k/a_{ij} \quad , \quad p_{ij}^0 = -a_{ij}^0/a_{ij}$$

These are elements of the field of fraction of the ring $O(U)$. (In such a case the singular locus $S(I)$ is exactly the set of poles of the p_{ij}^k's and the p_{ij}^0's.) Indeed by putting $u^0 = u$, $u^i = D_i u$ ($1 \leq i \leq n$) one is lead to replace $Iu = 0$ by the $(n+1)$-Pfaffian system $d\,{}^t(u^0,\ldots,u^n) = \Omega\,{}^t(u^0,\ldots,u^n)$ where

$$\Omega = \begin{bmatrix} 0 & dx^1 & \cdots & dx^n \\ \Sigma_m p_{1m}^0 dx^m & \Sigma_m p_{1m}^1 dx^m & \cdots & \Sigma_m p_{1m}^n dx^m \\ & & & \\ \Sigma_m p_{nm}^0 dx^m & \Sigma_m p_{nm}^1 dx^m & \cdots & \Sigma_m p_{nm}^n dx^m \end{bmatrix}.$$

Then the integrability condition $d\Omega - \Omega \wedge \Omega = 0$ clearly boils down to (7.4).

Remark: By using the equalities of the second line of (7.4), we see that when $n \geq 2$ the p_{ij}^0's are polynomials of degree 2 in the p_{ij}^k's ($1 \leq i,j,k \leq n$) and their first order derivatives.

With the previous notations, the system

(7.5) $P_{ij}' u = 0$ ($1 \leq i,j \leq n$)

is said to be <u>integrable</u> if the system (7.4) —its <u>integrability condition</u> — holds. There is a one to one correspondence between the set of integrable systems of the form (7.5) and the set of ideals I of $D(U)$ such that $r(I) = n+1$ and $I = \tilde{I}$.

§ 7.5 Differential Equations on Manifolds

Let X be a complex manifold and let $\{U_\alpha\}$ be a system of local charts consisting of open subsets of $\mathbb{C}^n$. A <u>differential equation</u> $\underline{E}$ on X is a system (I_α) of ideals I_α of $D(U_\alpha)$ such that on each open subset $U_\alpha \cap U_\beta$ the restrictions of I_α and of I_β to $U_\alpha \cap U_\beta$ coincide. We identify two differential equations $\underline{E}^j$ ($j = 1, 2$) defined by systems of ideals I_α^j of $D(U_\alpha^j)$ (where $\{U_\alpha^j\}$ are two systems of local charts) if the union of two systems of ideals define a differential equation on X. <u>The singular locus</u> $S(\underline{E})$ of $\underline{E}$ is the analytic subset of X which is the union of the singular loci of the I_α's. <u>The rank</u> $r(\underline{E})$ of $\underline{E}$ is defined to be the rank of any I_α. If $r(\underline{E}) < \infty$ then $\underline{E}$ is said to be <u>holonomic</u>. Let Y be an open submanifold of X then <u>the restriction</u> of an equation on X to Y and an <u>extention</u> of an equation on Y to X are defined in an obvious way.

Two complex manifolds X_1 and X_2 are called <u>bimeromorphic</u> if there is an subvariety G of $X_1 \times X_2$ such that the projections $p_j : G \to X_j$ ($j = 1, 2$) have the following properties: (i) p_j is proper, (ii) there are open subsets G_j' of G and X_j' of X_j such that $G - G_j'$ and $X_j - X_j'$ are contained in subvarieties of G and X_j, respectively; and that the restriction of p_j on G_j' gives a biholomorphic map $G_j' \to X_j'$.

Let X and Y be two complex manifolds which are bimeromorphic then there is a one to one correspondence between the set of differential equations $\underline{E} = (I_\alpha)$ on X such that $I_\alpha = \tilde{I}_\alpha$ and the set of differential equations $\underline{F} = (J_\alpha)$ on Y such that $J_\alpha = \tilde{J}_\alpha$. Indeed since there are open subsets X' of X and Y' of Y such that $X - X'$ and $Y - Y'$ are contained in subvarieties of X and Y, respectively, and that there is a biholomorphic map $X' \to Y'$, we have only to apply the discussions at the end of § 7.1.

Let $\underline{E}$ be one of the HGDE's : $P_1 u = 0$, $P_2 u = 0$ defined on $\mathbb{C}^2$. The singular locus in $\mathbb{C}^2$ of the equation $\underline{E}$ is the singular locus of the ideal $\tilde{I}$ where I is the ideal of $D(\mathbb{C}^2)$ generated by P_1 and P_2. Let X be a smooth compactification of $U_0 = \mathbb{C}^2$ obtained by patching open sets $U_1, \ldots, U_s$. Since the coefficients of the operators P_1 and P_2 are polynomials, there is an equation on X such that the restriction to U_0 coincides with I. The singular locus in X of the equation $\underline{E}$ is the singular locus of the equation $\underline{E} = (I_\alpha)$ on X such that $I_0 = \tilde{I}$ and $I_\alpha = \tilde{I}_\alpha$.

For a holonomic equation $\underline{E}$ on X, the monodromy representation $\rho: \Pi_1(X - S(\underline{E}), a) \to GL(r(\underline{E}), \mathbb{C})$ is defined, as in § 2.3, by using a system of linearly independent holomorphic solutions at $a \in X - S(\underline{E})$. The image of this representation is called the monodromy group. The conjugacy class of the monodromy representation and the monodromy group do not depend on the choice of a nor on the chosen system of linearly independent solutions. The projective monodromy group is defined as the image of the monodromy group by the natural map $GL(r(I), \mathbb{C}) \to PGL(r(I), \mathbb{C})$.

§ 7.6 Regular Singularities

Let U be an open subset of $\mathbb{C}^n$ and let I be a holonomic ideal of $D(U)$ of rank r such that $I = \tilde{I}$. Let x_0 be a non-singular point of $S(I)$. Choose a contractible open neighbourhood B of x_0 in U such that the hypersurface $B \cap S(I)$ is given by the zero locus of a holomorphic function f on B and that the fundamental group $\Pi_1(B - S(I) \cap B, x_0')$ is isomorphic to $\mathbb{Z}$, where x_0' is any point in $B - S(I) \cap B$. Let $u^1, \ldots, u^r$ be linearly independent holomorphic solutions of $Iu = 0$ at x_0' and let γ be a loop in $B - S(I) \cap B$ starting and ending at x_0' which represents a generator of $\Pi_1(B - S(I) \cap B, x_0')$. Let A be the <u>circuit matrix</u> of $(u^1, \ldots, u^r)$ along γ which is defined, as in the 1-dimensional case (see § 2.3), by using the analytic continuation. Let B be a matrix such that $\exp 2\pi i M = A$. If the single valued vector $\exp(M \log f)\,^t(u_1, \ldots, u_r)$ has at most poles along $S(I) \cap B$, then the point x_0 is said to be <u>regular singular</u>. The equation $Iu = 0$ is said to be <u>regular singular</u> in a domain $W \subset U$ if any non-singular point of $S(I)$ in W is regular singular.

Let X be a compact complex manifold obtained by patching open subsets U_α of $\mathbb{C}^n$. An equation $\underline{E} = (I_\alpha)$ on X is said to be <u>Fuchsian</u> if for all α the equation $I_\alpha u = 0$ on U_α is regular singular. If Y is another compact complex manifold bimeromorphically equivalent to X then the corresponding differential equation on Y is also Fuchsian. An equation $Iu = 0$ on $\mathbb{C}^n$ is said to be <u>Fuchsian</u> if there is an extension of the equation to the projective space P_n which is Fuchsian. (If the ideal I is generated by operators with polynomial coefficients then it is always extendable to P_n.)

Please check that all equations in § 6.1, ..., § 6.7 are Fuchsian.

Chapter 8 Schwarzian Derivatives II

In this chapter we define, as a generalization of the Schwarzian derivative introduced in Chapter 4, a system of $PGL(n+1,\mathbb{C})$-invariant operators S_{ij}^k ($1 \leq i,j,k \leq n$) on non-degenerate maps of n-variables, where we assume $n \geq 2$. These operators will be key tools for constructing uniformizing equations in Chapters 10 and 12.

§ 8.1 Definitions and Properties

In this section and the next one we calculate differentials formally so we do not worry about where functions and maps are defined. Let n be an integer greater than or equal to two and let $z : x = (x^1,..,x^n) \to z = (z^1,..,z^n)$ be a non-degenerate map; this means the Jacobian determinant does not vanish identically.

Here and in the sequel we use <u>the Einstein convention</u> i.e. if an index occurs twice in a term, once as a superscript and once as a subscript, we sum over that index from 1 through n.

With this convention, the Schwarzian derivatives $S_{ij}^k(z;x)$ ($i,j,k = 1,..,n$) of z with respect to x are

$$S_{ij}^k(z;x) = \frac{\partial^2 z^p}{\partial x^i \partial x^j} \frac{\partial x^k}{\partial z^p} - \frac{1}{n+1} \left(\delta_i^k \frac{\partial^2 z^p}{\partial x^q \partial x^j} \frac{\partial x^q}{\partial z^p} + \delta_j^k \frac{\partial^2 z^p}{\partial x^q \partial x^i} \frac{\partial x^q}{\partial z^p} \right)$$

Proposition: The Schwarzian derivatives have the following properties:

1) (PGL$(n+1,\mathbb{C})$-invariance) If $A \in \text{PGL}(n+1,\mathbb{C})$, then

$$S^k_{ij}(Az;x) = S^k_{ij}(z;x)$$

($1 \leq i,j,k \leq n$)

2) (the connection formula) If $y \mapsto x$ is a non-degenerate map, then

$$S^k_{ij}(z;y) = S^k_{ij}(x;y) + S^r_{pq}(z;x) \frac{\partial x^p}{\partial y^i} \frac{\partial x^q}{\partial y^j} \frac{\partial y^k}{\partial x^r}$$

($1 \leq i,j,k \leq n$)

Remarks: 1) By using the properties 1) and 2) one sees that $S^k_{ij}(z;x) = 0$ for $1 \leq i,j,k \leq n$ if and only if the map $x \mapsto z(x)$ is projective linear.

2) If $z(x)$ is PGL$(n+1,\mathbb{C})$-multivalued, i.e. if all the branches of $z(x)$ are projectively related to $z(x)$, then the $S^k_{ij}(z;x)$'s are single valued.

3) If we dare put $n = 1$, then by this definition of the derivative we have $S^1_{11}(z;x) = 0$ for any function $z(x)$; which does not give the Schwarzian derivative $\{z;x\}$ defined in Chapter 4 (see § 8.2 Remark).

Assertion 1) is proved in the next section, while assertion 2) can be verified in the same way as we did in § 4.3 using this time the proposition in § 8.3; this verification is left to the reader.

§ 8.2 Relations With Differential Equations

<u>Proposition</u>: Let $z(x)$ be a non-degenerate $PGL(n+1,\mathbb{C})$-multi-valued map and put $p_{ij}^k = S_{ij}^k(z;x)$ ($1 \leq i,j,k \leq n$). Then there are $n+1$ linearly independent solutions $u_0,\ldots,u_n$ of the system

$$D_i D_j u - p_{ij}^k D_k - p_{ij}^0 u = 0 \qquad (1 \leq i,j,k \leq n)$$

with single valued coefficients. Furthermore we have $z(x) = (u_1/u_0,\ldots,u_n/u_0)$.

We refer to such systems as <u>systems of the form</u> E which have already been studied in § 7.4 under the name of the form (7.5).

<u>Remark</u>: If a system of the form E is integrable (i.e. of rank $n+1$) and if $n \geq 2$ then the coefficients p_{ij}^0 can be expressed in terms of the p_{ij}^k's (with $k \neq 0$) and their derivatives (cf. § 7.4 and § 12.2). If $n = 1$ then as we saw in § 4.2, we have $p_{11}^1 = 0$ and $p_{11}^0 = \{z;x\}$.

For notational simplicity we denote by $(f)_j$ the partial derivative $D_j f$ of f with respect to x^j.

<u>Proof of the proposition</u> (cf. § 4.2): Put

$$u_0 = (\det (\partial z/\partial x))^{-1/(n+1)}$$

$$u_i = z^i(x) u_0 \qquad (1 \leq i \leq n)$$

and consider the following system of linear differential equations with $\mathbb{C}u_0 + \ldots + \mathbb{C}u_n$ as its solution space:

$$\begin{vmatrix} u_0 & \ldots & u_n & u \\ (u_0)_1 & \ldots & (u_n)_1 & (u)_1 \\ \vdots & & & \\ (u_0)_n & \ldots & (u_n)_n & (u)_n \\ (u_0)_{ij} & \ldots & (u_n)_{ij} & (u)_{ij} \end{vmatrix} = 0$$

A projective change

$$(z^1, \ldots, z^n) \mapsto \left(\frac{a_{10} + a_{1i}z^i}{a_{00} + a_{0i}z^i}, \ldots, \frac{a_{n0} + a_{ni}z^i}{a_{00} + a_{0i}z^i} \right)$$

with $(a_{ij}) \in SL(n+1, \mathbb{C})$ gives a linear change

$$\begin{bmatrix} u_0 \\ \vdots \\ u_n \end{bmatrix} \mapsto \begin{bmatrix} a_{00} & a_{01} & \ldots & a_{0n} \\ \vdots & \vdots & & \vdots \\ a_{n0} & a_{n1} & \ldots & a_{nn} \end{bmatrix} \begin{bmatrix} u_0 \\ \vdots \\ u_n \end{bmatrix}$$

Thus the coefficients of the differential equations, which are subdeterminants of the big matrix, are single valued. Let us compute these subdeterminants. The coefficient of $(u)_{ij}$ is

$$\begin{vmatrix} u_0 & \ldots & u_n \\ (u_0)_1 & \ldots & (u_n)_1 \\ & \ldots & \\ (u_0)_n & \ldots & (u_n)_n \end{vmatrix} = \begin{vmatrix} u_0 & z^1 u_0 & \ldots & z^n u_0 \\ (u_0)_1 & (z^1)_1 u_0 + z^1 (u_0)_1 & \ldots & (z^n)_1 u_0 + z^n (u_0)_1 \\ & \ldots & & \\ (u_0)_n & (z^1)_n u_0 + z^1 (u_0)_n & \ldots & (z^n)_n u_0 + z^n (u_0)_n \end{vmatrix}$$

$$= \begin{vmatrix} u_0 & 0 & \ldots & 0 \\ (u_0)_1 & z^1(u_0)_1 & \ldots & z^n(u_0)_1 \\ & \ldots & & \\ (u_0)_n & z^1(u_0)_n & \ldots & z^n(u_0)_n \end{vmatrix} = u_0^{n+1} \det(\partial z / \partial x) = 1$$

If the symbol $< k$ means that the k-th row is omitted, the coefficient of $(u)_k$ $(k = 1,..,n)$ is

$$(-1)^{n+1-k} \begin{vmatrix} u_0 & \cdots & u_n \\ (u_0)_1 & \cdots & (u_n)_1 \\ \cdots & \cdots & \cdots \\ (u_0)_n & \cdots & (u_n)_n \\ (u_0)_{ij} & \cdots & (u_n)_{ij} \end{vmatrix} < k$$

$$= (-1)^{n+1-k} \begin{vmatrix} u_0 & 0 & \cdots \\ (u_0)_1 & u_0(z^1)_1 & \cdots \\ \cdots & \cdots & \\ (u_0)_n & u_0(z^1)_n & \cdots \\ (u_0)_{ij} & u_0(z^1)_{ij} + (u_0)_i(z^1)_j + (u_0)_j(z^1)_i & \cdots \end{vmatrix} < k$$

$$= \begin{vmatrix} (z^1)_1 & \cdots \\ \vdots & \\ (z^1)_{k-1} & \cdots \\ (z^1)_{ij} + (z^1)_j(u_0)_i/u_0 + (z^1)_i(u_0)_j/u_0 & \cdots \\ (z^1)_{k+1} & \cdots \\ \vdots & \\ (z^1)_n & \cdots \end{vmatrix}$$

$$\times \begin{vmatrix} \partial x^1/\partial z^1 & \cdots & \partial x^n/\partial z^1 \\ \vdots & & \vdots \\ \partial x^1/\partial z^n & \cdots & \partial x^n/\partial z^n \end{vmatrix}$$

The last matrix comes from the term u_0^{n+1}, which is by definition equal to $(\det(\partial z/\partial x))^{-1} = \det(\partial x/\partial z)$. Thus the coefficient of $(u)_k$ is

$$\begin{vmatrix} 1 & & & & \\ & \ddots & & & \\ & & 1 & & \\ \cdots & \cdots & A & \cdots & \\ & & 1 & & \\ & & & \ddots & \\ & & & & 1 \end{vmatrix} = A$$

where

$$A = \{(z^p)_{ij} + (z^p)_j (u_0)_i/u_0 + (z^p)_i (u_0)_j/u_0\} \, \partial x^k/\partial z^p.$$

On the other hand we can easily see that

$$(u_0)_i/u_0 = -(n+1)^{-1}|\partial z/\partial x|^{-1} \, \partial|\partial z/\partial x|/\partial x^i$$

$$= -(n+1)^{-1}(z^p)_{qi} \, \partial x^q/\partial z^p.$$

This completes the proof.

§ 8.3 A Canonical Form of Differential Equations

<u>Definition</u>:   A system

$$E: \quad D_i D_j u = p_{ij}^k D_k u + p_{ij}^0 u \qquad (i,j = 1,\ldots,n)$$

is said to be in (projectively) <u>canonical form</u> if

$$p_{im}^m = 0 \qquad (i = 1,\ldots,n)$$

Check that the system in the proposition in § 8.2 is in canonical form. If E is integrable (see § 7.4), then by using $n+1$ linearly independent solutions $u_0, \ldots, u_n$, the Wronskian W is defined by

$$W = W(u) = \begin{vmatrix} u_0 & \cdots & u_n \\ (u_0)_1 & \cdots & (u_n)_1 \\ \vdots & & \vdots \\ (u_0)_n & \cdots & (u_n)_n \end{vmatrix}$$

<u>Remarks</u>: 1) If E is integrable, then E is in canonical form if and only if the Wronskian is constant, since we have

$$d \log W = p^m_{im} \, dx^i.$$

2) If $n = 1$, then this definition coincides with that in § 4.3.

<u>Definition</u>: The ratio of $n+1$ linearly independent solutions is called a <u>projective solution</u>. It is uniquely determined modulo $PGL(n+1, \mathbb{C})$. Its class is called <u>the</u> projective solution. Two systems are said to be <u>projectively equivalent</u> if their projective solutions are equal. Two ideals of rank $n+1$ of $D(U)$ for some open subset U in $\mathbb{C}^2$ are said to be <u>projectively equivalent</u> if the corresponding systems of the form E (see the end of § 7.4) are projectively equivalent.

<u>Proposition</u>: By replacing the unknown u by its product with a non-zero function of x, any integrable system of the form E is transformed into a uniquely defined system in canonical form without changing the projective solution. The coefficients are transformed as follows

$$p^k_{ij} \longrightarrow p^k_{ij} - \frac{1}{n+1} (\delta^k_i p^m_{jm} - \delta^k_j p^m_{im})$$

$$p^0_{ij} \longrightarrow p^0_{ij} - \frac{1}{n+1} ((p^m_{im})_j + p^m_{hm} p^h_{ij}) - (\frac{1}{n+1})^2 p^m_{im} p^h_{jh}$$

Proof: Analogous to that of the proposition in § 4.3. Instead of solving $a'/a + p/2 = 0$, to determine a, we have to solve the equation

$$d \log a = (n+1) \, p^m_{im} \, dx^i.$$

Since by using the integrability of $\underline{E}$ one sees that the exterior derivative of the right hand side is zero, and that the equation has a solution.

Remarks: 3) Even if one does not assume the integrability of $\underline{E}$, one can transform it into the canonical form by using the transformations of the coefficients above.
    4) If $n = 1$, this proposition reduces to that in § 4.3.
    4) The Schwarzian derivatives of a projective solution of $\underline{E}$ are the coefficients of the canonical form of $\underline{E}$.

§ 8.4 Projective Equations and Projective Connections

Let X be a complex manifold obtained by patching open subsets U_α of $\mathbb{C}^n$ ($n \geq 2$), and consider differential equation $\underline{E} = (I_\alpha)$ on X such that the rank of the ideal I_α of $D(U_\alpha)$ is $n+1$. In this section all ideals I_α are supposed to have the property $I_\alpha = \tilde{I}_\alpha$. Two differential equations $\underline{E}^1 = (I^1_\alpha)$ and $\underline{E}^2 = (I^2_\alpha)$ are said to be <u>projectively equivalent</u> if for all α the ideals I^1_α and I^2_α of $D(U_\alpha)$ are projectively equivalent. A class $(\underline{E})$ of projectively equivalent differential equations $\underline{E}$ is called a <u>projective equation</u> on X. Let $(\underline{E})$ be a projective equation on X, let $\underline{E} = (I_\alpha)$ be a representative of $(\underline{E})$, let E_α be the system of the form E which is uniquely determined (as in § 7.4) by I_α and let

$p_{\alpha ij}^{k}$ ($1 \leq i,j,k \leq n$) be the coefficients of the canonical form E_{α}^{0} of the system E_{α}. The coefficients $p_{\alpha ij}^{k}$ do not depend on the chosen equation $\underline{E}$. <u>The singular locus</u> $S((\underline{E}))$ of a projective equation $(\underline{E})$ on X is the analytic subset of codimension 1 in X which is the union of the set of poles of the coefficients $p_{\alpha ij}^{k}$'s. (Notice that the system (I_{α}^{0}) of ideals I_{α}^{0} of $D(U_{\alpha})$ corresponding to the system E_{α}^{0} does not define a differential equation on X.) <u>The projective solution</u> of $(\underline{E})$ is defined in an obvious way. A projective equation is called <u>Fuchsian</u> if any system E_{α}^{0} is regular singular along $S((\underline{E})) \cap U_{\alpha}$.

On the other hand a system $\{p_{\alpha ij}^{k}\}$ of meromorphic (resp. holomorphic) functions $p_{\alpha ij}^{k}$ ($1 \leq i,j,k \leq n$) on U_{α} satisfying the symmetry conditions $p_{\alpha ij}^{k} = p_{\alpha ji}^{k}$ and the equalities

$$(8.1) \quad p_{\alpha ij}^{k} - p_{\beta pq}^{r} \frac{\partial x_{\beta}^{p}}{\partial x_{\alpha}^{i}} \frac{\partial x_{\beta}^{q}}{\partial x_{\alpha}^{j}} \frac{\partial x_{\alpha}^{k}}{\partial x_{\beta}^{r}} = S_{ij}^{k}(x_{\beta};x_{\alpha})$$

is called a <u>meromorphic</u> (resp. <u>holomorphic</u>) <u>projective connection</u>, where $x_{\alpha} = (x_{\alpha}^{1},\ldots,x_{\alpha}^{n})$ is a coordinate system of U_{α} (cf. [Ko-O]). A meromorphic projective connection $\{p_{\alpha ij}^{k}\}$ is called <u>flat</u> if for each α there exists an integrable system of the form E with coefficients $p_{ij}^{k} = p_{\alpha ij}^{k}$ ($1 \leq i,j,k \leq n$). Notice that if we substitute

$$x = x_{\beta}, \quad y = x_{\alpha}, \quad S_{ij}^{k}(z;y) = p_{\alpha ij}^{k} \quad \text{and} \quad S_{ij}^{k}(z;x) = p_{\beta ij}^{k}$$

in the connection formulae of the Schwarzian derivatives in § 8.1 then the formulae turn out to be the equalities (8.1). It implies that there is an one to one correspondence between the set of projective equations on X and the set of flat meromorphic projective connections on X. If the manifold X admits a flat holomorphic projective connection then, by changing the coordinates $x_\alpha = (x_\alpha^1, \ldots, x_\alpha^n)$ into a projective solution $z_\alpha = (z_\alpha^1, \ldots, z_\alpha^n)$ of the equation E_α of the form E with coefficients $p_{ij}^k = p_{\alpha ij}^k$ for each α, one gets a holomorphic projective structure i.e. a system of local coordinates such that the the coordinate changes are given by projective transformations.

Remark: From the view point of the theory of differential equations we are now working, non-integrable systems are not interesting. If for example the rank of an equation of the form E is zero then the system is equivalent to the equation " u = 0 " so that the coefficients p_{ij}^k have no special meaning. But in the theory of connections, projective connections which are not necessarily flat do make sense and are useful. Indeed from the existence of a holomorphic projective connection on X one can derive very strong results (see [Ko-O]).

§ 8.5 Local Properties of Schwarzian Derivatives

Definition: A map $x \mapsto z = z(x)$ is said to be ramified with exponents s along the hypersurface $\{x^1 = 0\}$ if $z(x)$ has the following expression

$$z^1(x) = (x^1)^s v^1, \qquad z^j(x) = v^j \quad (j = 2, \ldots, n)$$
$$\det(\partial z/\partial x) = (x^1)^{s-1} u$$

where the v^j's ($1 \le j \le n$) and u are holomorphic functions not divisible by x^1.

Proposition: If $z = z(x)$ is ramified along $\{x^1 = 0\}$ with exponent s then

$$S_{ij}^k(z;x), \quad S_{i1}^1(z;x), \quad x^1 S_{11}^k(z;x), \quad (x^1)^{-1} S_{ij}^1(z;x)$$

and $\quad S_{11}^1(z;x) - \dfrac{s-1}{n+1} \dfrac{1}{x^1}$

are holomorphic for $2 \le i, j, k \le n$.

Proof: We prove only when $n = 2$. By substituting the following expressions

$$\begin{bmatrix} \dfrac{\partial z^1}{\partial x^1} & \dfrac{\partial z^1}{\partial x^2} \\ \dfrac{\partial z^2}{\partial x^1} & \dfrac{\partial z^2}{\partial x^2} \end{bmatrix} = \begin{bmatrix} s(x^1)^{s-1} v^1 + (x^1)^s \dfrac{\partial v^1}{\partial x^1} & (x^1)^s \dfrac{\partial v^1}{\partial x^2} \\ \dfrac{\partial v^2}{\partial x^1} & \dfrac{\partial v^2}{\partial x^2} \end{bmatrix}$$

$$\begin{bmatrix} \dfrac{\partial x^1}{\partial z^1} & \dfrac{\partial x^1}{\partial z^2} \\ \dfrac{\partial x^2}{\partial z^1} & \dfrac{\partial x^2}{\partial z^2} \end{bmatrix} = \begin{bmatrix} (x^1)^{s-1} \dfrac{\partial v^2}{\partial x^2} & -(x^1)^s \dfrac{\partial v^1}{\partial x^2} \\ (x^1)^{1-s} \dfrac{\partial v^2}{\partial x^2} & sv^1 + x^1 \dfrac{\partial v^1}{\partial x^1} \end{bmatrix} \times u^{-1}$$

into the definition of the Schwarzian derivatives we have

$$S_{11}^1 = \{ s(s-1)(x^1)^{s-2} v^1 + 2s(x^1)^{s-1} \dfrac{\partial v^1}{\partial x^1} + (x^1)^s \dfrac{\partial^2 v^1}{\partial (x^1)^2} \}$$

$$\times (x^1)^{1-s} \dfrac{\partial v^2}{\partial x^2} \dfrac{\partial^2 v^2}{\partial (x^1)^2} \{ -x^1 \dfrac{\partial v^1}{\partial x^2} \} u^{-1}$$

$$- \dfrac{2}{3}(s-1)(x^1)^{-1} - \dfrac{2}{3} \dfrac{\partial}{\partial x^1} \log u$$

$$= \{s(s-1) v^1 \dfrac{\partial v^2}{\partial x^2} u^{-1} - \dfrac{2}{3}(s-1)\}(x^1)^{-1} + H_1.$$

Since we have

$$\det(\partial z/\partial x) = \{s(x^1)^{s-1}v^1 + (x^1)^s \frac{\partial v^1}{\partial x^1}\} \frac{\partial v^2}{\partial x^2} - (x^1)^s \frac{\partial v^1}{\partial x^2} \frac{\partial v^2}{\partial x^1}$$

$$= (x^1)^{s-1}\{sv^1 \frac{\partial v^2}{\partial x^2} + x^1 H_2\}$$

by the definition of the function u, we have

$$u^{-1} = (sv^1 \frac{\partial v^2}{\partial x^2})^{-1} + x^1 H_3.$$

In what precedes H_1, H_2 and H_3 denotes holomorphic functions. By substituting the last expression of u^{-1} into the above expression of S^1_{11} we have the desired result for S^1_{11}. By following the same idea the remaining assertions can be proved much easily than the present one.

Definition: A system of the form E is said to have <u>ramifying singularities of exponent</u> s <u>along</u> $\{x^1 = 0\}$ if the coefficients p^k_{ij} of the canonical form have the following properties:

$$p^k_{ij}, \quad p^1_{i1}, \quad x^1 p^k_{11}, \quad (x^1)^{-1} p^1_{ij} \quad \text{and} \quad p^1_{11} - \frac{s-1}{n+1} \frac{1}{x^1}$$

are holomorphic for $2 \leq i,j,k \leq n$.

Chapter 9 The Riemann and Riemann-Hilbert Problems II

In this chapter we discuss the Riemann problem and the question of accessory parameters in n variables. As we saw in Chapter 3, the Riemann problem is simple if n = 1. However we shall see now that it is very complicated if $n \geq 2$; for several variables the Riemann-Hilbert problem forms part of the Riemann problem. To counterbalance this extra complexity, we find that in several variables the question of accessory parameters turns out to be simple, at least in the cases we look at.

§ 9.1 The Riemann Problem in Several Variables

Let $\underline{E}$ be a Fuchsian equation of rank r on a complex manifold X of dimension n with only regular singularities along a hypersurface $S \subset X$ and let $\{S_j\}$ be the set of irreducible components of S. Since in n (≥ 2) variables we have no "indicial equations" (§ 2.5), we use a slightly different method to describe the local behaviour of the solutions. Denote by $u_1, \ldots, u_r$ a set of linearly independent holomorphic solutions of $\underline{E}$ at a fixed point $a \in X - S$. Let μ_j be a loop in X - S starting at a and defined as follows. Follow a path to a point of S_j at where S is non-singular, then go once around the hypersurface S_j in the positive direction and return to a along the same path as used on the outward journey. (Note that it makes sense to refer to going around S_j in the positive direction.) Finally denote by $M_j \in GL(n, \mathbb{C})$ the circuit matrix of $(u_1, \ldots, u_r)$

along μ_j. The system of conjugacy classes $[\mu_j]$ of the μ_j's in the fundamental group $\Pi_1(X-S,a)$ and the conjugacy classes $[M_j]$ of the M_j's in $GL(r,\mathbb{C})$ (resp. in $PGL(r,\mathbb{C})$) is called the local behaviour (resp. the projective local behaviour) of the equation E.

Let X, S, S_j, a ε X-S and μ_j as above. A system L consisting of the classes $[\mu_j]$ and an arbitrary set of conjugacy classes N_j in $GL(r,\mathbb{C})$ (resp. in $PGL(n,\mathbb{C})$) is called an r-local (resp. projective r-local) datum on X along S. One can now pose

The Riemann problem: For a given r- local (resp. projective r-local) datum L on X along S, can we find a Fuchsian differential equation of rank r on X with only regular singularities along S such that its local behaviour (resp. its projective local behaviour) is L ?

Remark: In § 3.1 we posed the Riemann problem in one dimension for Fuchsian ordinary differential equations of rank 2 defined on $X = \mathbb{P}_1$ with regular singularities in $S = \{x_1,\ldots,x_{m+1}\}$. The set of eigenvalues of a circuit matrix M_j at x_j is the pair $s_j^{(1)}$, $s_j^{(2)}$ of characteristic exponents at x_j modulo $\mathbb{Z}$. (i.e. the set of the indicial equation § 2.5.) Thus if all the M_j's are semi-simple, the present Riemann problem reduces to the countably many Riemann problems posed in § 3.1, which we can solve. When $n \geq 2$, as the equation must be of rank r, we have to find meromorphic solutions of non-linear differential equations (namely the integrability conditions (7.3)) in which the unknowns are the coefficients of the equation we are seeking. For example, if $r = n+1$ we face to solve the system (7.4) in unknowns p_{ij}^k and p_{ij}^0 ($1 \leq i,j,k \leq n$). Unfortunately we have as yet no tool to study these non-linear differential equations directly.

To attack the Riemann problem, let us divide the problem into two parts in the hope that we can solve at least one of them.

1) <u>The problem of representations of</u> $\Pi_1(X-S,a)$: For a given r-local (resp. projective r-local) datum $\underline{L} = ([\mu_j], N_j)_j$ on X along S, can we find a homomorphism.

$$\rho : \Pi_1(X-S,a) \to GL(n,\mathbb{C}) \quad (\text{resp. } PGL(r,\mathbb{C}))$$

such that $\rho(\mu_j) \in N_j$?

2) <u>The Riemann-Hilbert problem</u> : For a given homomorphism

$$\rho : \Pi_1(X-S,a) \to GL(n,\mathbb{C}) \quad (\text{resp. } PGL(r,\mathbb{C})),$$

can we find a Fuchsian differential equation $\underline{E}$ on X whose monodromy representation (resp. projective monodromy representation) is ρ ?

If we allow extra regular singularities outside S, these additional singularities will lie on some hypersurface S' of X, the Riemann-Hilbert problem can then be solved as in Chapter 3. The method used there can be generalized to several variables without difficulty if we do not assume any estimate for S'. In case n = 1, putting $S = \{x_1, \ldots, x_{m+1}\}$, the fundamental group $\Pi_1(X-S,a)$ is a free group on m free generators. So if the local data around the x_j's satisfy only one condition (cf. the Fuchs relation in § 2.6), then we have an r-dimensional representation of the fundamental group. Hence in this case the first problem is easy. But if $n \geq 2$ then the structure of $\Pi_1(X-S,a)$ is in general very complicated and as yet we have no sufficient conditions on the local data for obtaining non-trivial representations of the fundamenatal group. (Some necessary conditions are known, cf. [Oht 1]).

Therefore in this book we do not develop this tactics of splitting up the Riemann problem and solving problems (1) and (2). After preparing the appropriate tools in Chapters 10 and 11, we give in Chapter 12 another method for solving the projective version of the Riemann problem in a very special case. The equations obtained by this method will be called the uniformizing equations of orbifolds; their projective monodromy groups are discrete subgroups of certain linear Lie groups which are the groups of automorphisms of symmetric spaces.

§ 9.2 Accessory Parameters

It seems natural that every aspect of the theory (structure of $\Pi_1(X-S,a)$, integrability conditions, etc) should become more complicated and harder to unravel when one passes from one variable to several variables. However one crucial difficulty, that of " accessory parameters ", can sometimes vanish in the case of several variables.

Let X be a compact complex manifold obtained by patching open subsets U_α of $\mathbb{C}^n$ and let $\underline{E}$ be a Fuchsian differential equation on X with only regular singularities along $S \subset X$. The space $AP(\underline{E})$ of accessory parameters of the equation $\underline{E}$ is the set of Fuchsian equations on X with only regular singularities along $S \subset X$ and with the same local behaviour as that of $\underline{E}$. The space $AP((\underline{E}))$ of accessory parameters of the projective equation $(\underline{E})$ is the set of Fuchsian projective equations (§ 7.6) on X with only regular singularities along $S \subset X$ and with the same projective local behaviour as that of $(\underline{E})$. In general we know little about the structure of $AP(\underline{E})$ and of $AP((\underline{E}))$; we saw in § 3.2 that in the 1-dimensional case if $X = \mathbb{P}_1$, $r = 2$ and if all the circuit matrices M_j are semi-simple then $AP(\underline{E})$ as well as $AP((\underline{E}))$ is the union of a countable number of copies of complex linear space of dimension #S - 3.

Let (X,S,b) be an n (≥ 2)-dimensional orbifold uniformized by the n-ball

$$B_n = \{(z_0,\ldots,z_n) \in P_n | \ |z_0|^2 - \Sigma |z_j|^2 > 0\}$$

and suppose further that the base space X is a projective manifold, that is, it can be embedded in a projective space of some dimension. As we shall see in the next chapter (§ 10.4), there exists a unique projective equation (E) on X of rank n+1 with regular singularities along S with the following properties. Firstly, its projective solution gives rise to the developing map $X \to B_n \subset P_n$. Secondly, its projective monodromy group Γ, which can also be identified with the covering transformation group of the cover $B_n \to X - \cup \{S_j | b_j = \infty\}$, is a lattice of $\text{Aut}(B_2)$. (A lattice is a discrete subgroup Γ of $\text{Aut}(B_n)$ such that the quotient space B_n/Γ has a finite volume with respect to the $\text{Aut}(B_n)$-invariant metric of B_n.) Thirdly, its projective local behaviour $([\mu_j],[M_j])_j$ has the property that $M_j^{b_j} = 1$ in $\text{PGL}(n,\mathbb{C})$. The projective equation (E) will be called <u>the uniformizing differential equation of the orbifold</u> (X,S,b).

The manifold X can be considered as a submanifold of the m-dimensional projective space P_m for some $m \geq n$. Let $(y^0, y^1, \ldots, y^m)$ be a system of homogeneous coordinates on P_m such that the projection $P_m - \{y^0 = 0\} \cong \mathbb{C}^m \to \mathbb{C}^n$ given by $(x^1,\ldots,x^n,\ldots,x^m) \mapsto (x^1,\ldots,x^n)$, where $x^i = y^i/y^0$ ($1 \leq i \leq n$), induces a surjective map $p: X \cap \{y^0 \neq 0\} \to \mathbb{C}^n$. There is an open subset U of $X \cap \{y^0 \neq 0\}$ such that U is mapped by p biholomorphically onto $p(U) \subset \mathbb{C}^n$. The manifold X is covered by finite number of such U's, say $\{U_\alpha\}$, and the U_α's are patched by the restriction of rational functions in $(x^1,\ldots, x^m)$ to $X \cap \{y^0 \neq 0\}$.

Fix an index β and let $\underline{E} = (I_\alpha)$ be the representative of the projective equation $(\underline{E})$ such that the system E_β of the form E corresponding to the ideal $I_\beta \subset D(U_\alpha)$ is in canonical form. Since E_β has ramifying singularities along $S(\underline{E}) \cap U_\beta$, in view of the proposition in § 8.5, the coefficients p_{ij}^k ($1 \le i,j,k \le n$) of E_β have simple poles along $S(\underline{E}) \cap U_\beta$. If $\underline{E}' = (I'_\alpha)$ is an equation representing an element of $AP((\underline{E}))$, the coefficients p'^k_{ij} ($1 \le i,j,k \le n$) of the system E'_β of the form E corresponding to the ideal $I'_\beta \subset D(U_\beta)$ are the restriction to U_β of rational functions in $x^1,\ldots,x^m$ whose poles in U_β lie on $S(\underline{E}) \cap U_\beta$.

We define a subset $AP_N((\underline{E}))$ of $AP((\underline{E}))$ as follows. Take those equations $(\underline{E}')$ of $AP((\underline{E}))$ such that when the system E'_β is put into canonical form on U_β the coefficients of this canonical form are the restriction of rational functions in $x^1,\ldots,x^m$ whose numerators and denominators have total degrees at most N. All such equations constitute a set $AP_N((\underline{E}))$ which is the union of at most countably many disjoint algebraic sets. The space $AP((\underline{E}))$ is the inductive limit of the inclusion maps $AP_N((\underline{E})) \to AP_{N+1}((\underline{E}))$ ($N = 1,2,\ldots$) and $AP((\underline{E}))$ has a natural topology as the inductive limit of the $AP_N((\underline{E}))$'s. Since the local monodromy and integrability give algebraic conditions on the space of coefficients, the point $(\underline{E}) \in AP((\underline{E}))$ has a neighbourhood in this topology which is an algebraic set. We assert that the point $(\underline{E})$ is isolated in the topological space $AP((\underline{E}))$.

Theorem: Let $(\underline{E})$ be the uniformizing differential equation of an orbifold (X,S,b) of dimenssion n uniformized by the ball B_n and suppose that X is a projective algebraic manifold. If $n \ge 2$ then the point $(\underline{E})$ is isolated in $AP((\underline{E}))$.

Proof: Let H be the fundamental group of $X-S$ with base point a, and $H(\mu^b)$ be the normal subgroup defined in § 5.1. Let $\Gamma \subset \text{Aut}(B_n) \cong PU(n,1)$ be the covering transformation group of the covering $B_n \to X$. Suppose there is a continuous curve $\{(\underline{E}(t))\}$ in $AP((\underline{E}))$ such that $(\underline{E}(0)) = (\underline{E})$. Let $U(t)$ be a projective solution of $(\underline{E}(t))$ at a, depending continuously on t such that $U(0)$ gives rise to the developing map $X \to B_n$. The projective monodromy representation of the projective equation $(\underline{E}(t))$ with respect to the solution $U(t)$ gives the homomorphism

$$M(t): H \to PGL(n+1,\mathbb{C}).$$

For $t = 0$ we have

$$M(0): H \to \Gamma \hookrightarrow PU(n,1) \hookrightarrow PGL(n+1,\mathbb{C})$$

On account of the fixed local data of $U(t)$ along the S_j's, $M(t)$ is trivial on $H(\mu^b)$. Recalling the isomorphism $\Gamma \cong H/H(\mu^b)$ in § 5.1, we see that $M(t)$ induces a representation $M'(t): \Gamma \to PGL(n+1,\mathbb{C})$. On the other hand we have the following <u>generalized rigidity theorem</u> due to Goldman, Johnson and Millson.

"<u>Theorem</u>([Gol],[J-M]): Let Γ be a lattice in $PU(n,1)$ where $n \geq 2$. Then any continuous deformation

$$m_t: \Gamma \to PGL(n+1,\mathbb{C})$$

(where m_0 is the inclusion $\Gamma \hookrightarrow PU(n,1) \hookrightarrow PGL(n+1,\mathbb{C})$) is a trivial deformation, that is, of the form $m_t(x) = g_t m_0(x) g_t^{-1}$ where g_t is a path in $PGL(n+1,\mathbb{C})$."

Apply this theorm by putting $m_t = M'(t)$ and denote by g_t the resulting path in $PGL(n+1,\mathbb{C})$. We know that the product $U(t)g_t$ gives rise to a developing map $X \to B_n$. This implies $(\underline{E}(0))$ and $(\underline{E}(t))$ have the same projective solution and so the theorem is proved.

<u>Remarks</u>: 1) As we saw in the proof, the space of accessory parameters of a uniformizing equations has a deep relation with the deformation (or the rigidity) of lattices.

2) It is natural to conjecture that $AP((\underline{E}))$ itself is discrete. Some examples (see Chapter 11 and 13) support this conjecture.

3) The original <u>rigidity theorem</u> due to Weil and the <u>strong rigidity theorem</u> due to Mostow affirm the (strong) rigidity of a lattice of a semi-simple Lie group G, having no $PU(1,1)$-factor, <u>in</u> G.

Chapter 10 The Gauss-Schwarz Theory in Two Variables

Let us recall the 1-dimensional case (Chapter 5) where orbifolds (X,S,b) on $X = \mathbb{P}_1$ with $S = \{x_1,\ldots,x_{m+1}\}$ and $b = (b_1,\ldots,b_{m+1})$ are concerned. If $m = 0$ or $m = 1$ and $b_1 \neq b_2$, then such orbifolds are not uniformizable. If $m = 1$ and $b_1 = b_2 = b$, then they are uniformizable and their developing maps are the b-th root or logarithm functions. If $m = 2$, then as we saw in § 5.3, it is always uniformizable and the universal uniformization is one of the three spaces: $M = \mathbb{P}_1$, $\mathbb{C}$ and $\mathbb{H}$, all of which are symmetric spaces. (See the note below.) It was this case that we encountered the HGDE and the Gauss-Schwarz theory.

We now look at the 2-dimensional case. We first discuss in § 10.1 the uniformizability of orbifolds and then in § 10.2 we give some methods to find orbifolds which are uniformized by the symmetric spaces $M = \mathbb{C}^2$, $\mathbb{P}_2$, $\mathbb{B}_2$, $\mathbb{P}_1 \times \mathbb{P}_1$ and $\mathbb{H} \times \mathbb{H}$. By applying these methods to orbifolds attached to certain line arrangements in $\mathbb{P}_2$, we get in § 10.3 some classical examples (studied by Picard, Terada, Deligne and Mostow) and also some new examples (found by Hirzebruch and Höfer) which are uniformized by $\mathbb{B}_2$. Next in § 10.4 we show the existence of uniformizing differential equations on the orbifolds uniformized by these symmetric spaces. The equations change their forms according to whether they are uniformized by $\mathbb{C}^2$, $\mathbb{P}_2$ and $\mathbb{B}_2$ or by $\mathbb{P}_1 \times \mathbb{P}_1$ and $\mathbb{H} \times \mathbb{H}$. The uniformizing equations of the classical examples turn out to be Appell's equations $F_1(a,b,b';c)$'s for some rational numbers a, b, b' and c, which are studied in § 10.5, while the uniformizing equations of the new examples are studied in Chapter 12.

Group theoretic facts used in this chapter and in Chapter 12 are collected in Chapter 11.

Note: A connected complex manifold M with a Hermitian metric is called a symmetric Hermitian space (in short, a symmetric space) if for each point p of M there exists an isometric and biholomorphic transformation of M onto M that is of order 2 and has p as an isolated fixed point. A symmetric space M is said to be irreducible if M is not the product of two symmetric spaces. The irreducible symmetric spaces of dimension 1 are $\mathbb{C}$, $\mathbb{P}_1$ and the upper half plane $\mathbb{H}$; those of dimension 2 are $\mathbb{P}_2$ and the unit ball $\mathbb{B}_2$; those of dimension 3 are $\mathbb{P}_3$, $\mathbb{B}_3$, quadratic hypersurface in $\mathbb{P}_4$ and Siegel's upper half space of genus 2. In this book, we do not refer the last two spaces.

§ 10.1 Uniformizability of Orbifolds

Since we have restricted ourselves to discussing non-singular uniformizations, it is more natural to consider orbifolds (X,S,b) on a rational surface X such that X - S is biholomorphic to the complement of a curve A in $\mathbb{P}_2$. We shall refer to orbifolds with this property as orbifolds over $\mathbb{P}_2$. When is an orbifold (X,S,b) over $\mathbb{P}_2$ uniformizable ? This is a fundamental question. We reformulate this question in the following way.

Question: Can we find to every curve A in $\mathbf{P}_2$, a uniformizable orbifold (X,S,b) such that $X - S = \mathbf{P}_2 - A$?

To give a general answer is beyond our reach. If A is the union of lines (referred to as a <u>line arrangement</u>), then there is a sufficient condition due to Kato (see also [Nam 1]). A multiple point of A is said to be <u>double</u> if it has multiplicity equal to 2, and is said to be <u>nondouble</u> if it has multiplicity greater than 2.

<u>Theorem</u> ([Kat 1]): Given a line arrangement $A = \{L_j\}_{j=1}^k$. If there is a nondouble multiple point on each line L_j ($1 \le j \le k$) then there is a uniformizable orbifold (X,S,b) such that $X - S = \mathbf{P}_2 - A$. Such an orbifold (X,S,b) can be constructed as follows: Blow up the nondouble multiple points of A to get a surface X, in which we have a configuration S of curves: the strict transforms of the L_j's (which we also call L_j) and the exceptional curves obtained by blowing up the nondouble multiple points. Give <u>any</u> weight b_j to the L_j. For an exceptional curve which is the blowing up of a singular multiple point $p = A_{j_1} \cap \ldots \cap A_{j_r}$ ($r \ge 3$), attach the least common multiple of the following numbers

$\{b_{j_i} | A_{j_i}$ has a nondouble multiple point other than p$\}$.

Then the orbifold (X,S,b) thus obtained admits a <u>finite</u> uniformization.

<u>Remarks</u>: 1) In brief, this theorem says that if one assumes certain complexity condition on A (existence of nondouble multiple points), then without any extra restrictions on the weights of lines L_j (weights b_j are arbitrary), the orbifold X will be uniformizable. In this sense, it can be considered

as a partial affirmative answer to the 2-dimensional analogue of the Fenchel conjecture (see § 5.1)

2) For a given line arrangement A, the method above is <u>not</u> the unique way to obtain a uniformizable orbifold (X,S,b) such that $X - S = P_2 - A$.

In contrust to the 1-dimensional case, when we look at dimension two we know almost nothing about simply connected complex surfaces except for symmetric spaces. For this reason we only consider uniformizable orbifolds whose universal uniformizations are P_2, $\mathbb{C}^2$, the unit ball

$$B_2 = \{(z^1, z^2) \in \mathbb{C}^2 \mid |z^1|^2 + |z^2|^2 < 1\},$$

$P_1 \times P_1$ and $H \times H$, which are the most important of the two dimensional symmetric spaces.

§ 10.2 Orbifolds Whose Universal Uniformizations are Symmetric Spaces

How can we find orbifolds (X,S,b) whose universal uniformizations M are the symmetric spaces $\mathbb{C}^2$, P_2, B_2, $P_1 \times P_1$ and $H \times H$? In this paragraph, the orbifolds are not assumed to be over P_2.

(1) <u>The case</u> $M = \mathbb{C}^2$. Our problem reduces to finding discrete subgroups of $E(2) = U(2) \ltimes \mathbb{C}^2$ such that the quotients of M, by these groups, are non-singular. Such subgroups are classified in [T-Y] and [K-T-Y]. Here is the complete list of orbifolds on $X = \mathbb{P}_2$ which are uniformized by $\mathbb{C}^2$. The numeral beside a curve denotes the weight b of the curve.

i) A conic and 3 tangents:

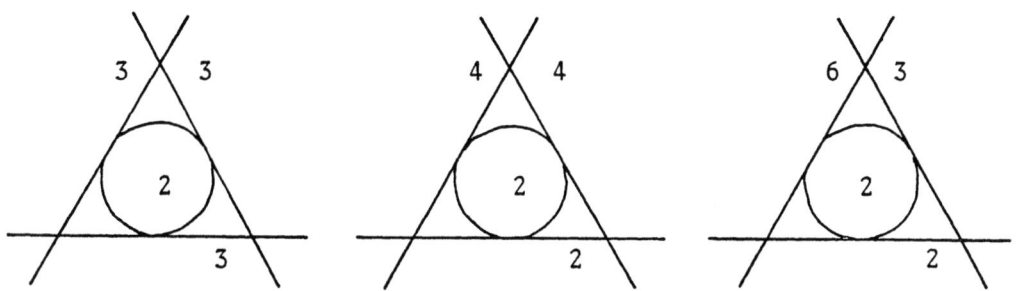

ii) Complete quadrilateral: iii) A conic and 4 tangents:

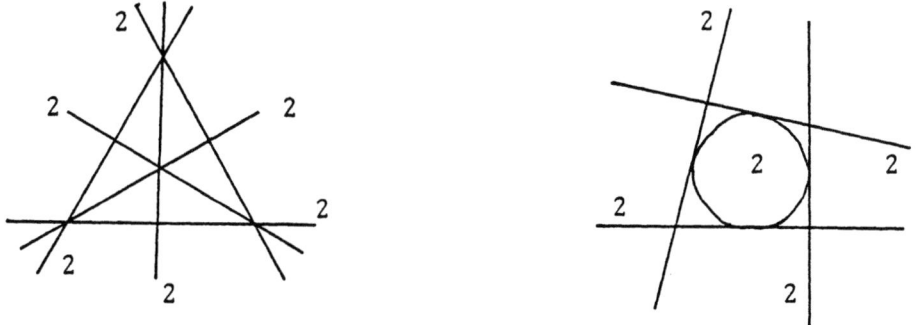

iv) Elliptic curve with 9 cusps, weight = 2

(2) <u>The case</u> $M = \mathbb{P}_2$ <u>and</u> $\mathbb{B}_2$. For the case $M = \mathbb{P}_2$, the orbifolds in question are the quotients of $\mathbb{P}_2$ by finite subgroups of $\text{Aut}(\mathbb{P}_2) \cong \text{PGL}(3,\mathbb{C})$. Although there are too many lattices of $\text{Aut}(\mathbb{B}_2) \cong \text{PU}(2,1)$ to be classified, we have the

following criterion for deciding whether an orbifold is a ball quotient. Let Y be a smooth projective algebraic surface. The Chern numbers $c_1(Y)^2$ and $c_2(Y)$ are defined. The divisor of any meromorphic 2-form on Y (canonical divisor) defines the canonical class $K(Y)$. The number $c_1(Y)^2$ is equal to the self-intersection number $K(Y)^2$ of the canonical divisor. The number $c_2(Y)$ is nothing but the Euler-Poincaré characteristic $e(Y)$ of Y. Hirzebruch proportionality (cf. [Hir 5]) tells us that if Y has P_2 or B_2 as the universal covering manifold, then the equality

$$c_1(Y)^2 = 3c_2(Y)$$

holds. Conversely we have,

Theorem ([Miy 2], [Yau]): If for a surface Y we have $c_1(Y)^2 = 3c_2(Y) > 0$ then the universal cover M of Y is either P_2 or B_2.

Remark : 1) Under the hypothesis of the theorem, several sufficient conditions are known for Y to have its universal cover $M = B_2$. For example if $K(Y) \cdot C > 0$ for all curves $C \subset Y$ then $M = B_2$.

This celebrated theorem generalizes to orbifolds (X,S,b). For simplicity, we assume that each S_j is a non-singular curve and that $S = \cup S_j$ has only double multiple points. Let us define the Euler-Poincaré characteristic $e(X,b) = c_2(X,b)$ and the canonical divisor $K(X,b) = c_1(X,b)$ of the orbifolds (X,S,b) as follows:

$$c_2(X,b) = e(X) + \Sigma_j (\frac{1}{b_j} - 1)(e(S_j) - \tau_j) + \Sigma_p (\frac{1}{b_i b_j} - 1)$$

where $e(S_j)$ is the Euler-Poincaré characteristic of the curve S_j, τ_j is the cardinality of $S_j \cap S$, and the summation Σ_p is over the nondouble points of S; and

$$c_1(X,b) = K(X) + \Sigma_j (1 - \frac{1}{b_j})S_j$$

respectively. Then we have

<u>Theorem</u> ([Ko R]): If for an orbifold (X,S,b) we have $(c_1(X,b))^2 = 3c_2(X,b) > 0$ then (X,S,b) is uniformizable and the universal uniformization M is either $\mathbf{P}_2$ or $\mathbf{B}_2$.

<u>Remarks</u>: 2) Under the hypothesis of the theorem, several sufficient conditions are known for (X,S,b) to have the uniformization $M = \mathbf{B}_2$. We do not discuss them here firstly because they need lengthy preparation and secondly because, for the concrete examples we are going to give in the next paragraph, we can easily see that their universal uniformizations cannot be $\mathbf{P}_2$ by showing that their covering groups are infinite (§ 10.3 Remark 3).
    3) If the orbifold (X,S,b) admits a finite uniformization Y, then letting d be the sheet number of the projection $p: Y \to X$, we can show that

$$c_2(Y) = d\, c_2(X,b), \quad c_1(Y) = p^* c_1(X,b)$$

and $\quad c_1(Y)^2 = d\, c_1(X,b)^2$.

Thus $c_1(X,b)^2 = 3c_2(X,b)$ implies $c_1(Y)^2 = 3c_2(Y)$. But we have no systematic method, in general, to construct (to prove the existence of) a finite uniformization of a given orbifold satisfying the condition of the theorem. This theorem is powerful because it does not assume the existence of a finite uniformization rather it implies the existence (cf. Selberg's Theorem in § 5.1).

4) Although we assumed that S admits only normal crossings, the theorem can be generalized for any S. But when an orbifold (X,S,b) is uniformized by the ball then S admits only very restricted types of singularities and the weights b are also restricted (see § 11.2 and 11.3).

(3) <u>The case</u> $M = P_1 \times P_1$ <u>and</u> $H \times H$. If a surface Y is covered by $P_1 \times P_1$ or H×H then the proportionality tells us that

$$c_1(Y)^2 = 2c_2(Y).$$

But in this case, this equality does not characterize such surfaces. (Counter examples are given in [K-N].) Nevertheless, a sufficient condition for orbifolds to be uniformized by $H \times H$, is known ([K-N]). The following is an example of a quotient of H×H which is due to Hirzebruch([Hir 4]).

<u>Orbifold</u> $X(\sqrt{2})$: Take the base space X to be P_2 and the singular locus S to be a curve of degree ten consisting of four lines and three conics, with all weights equal to 2. In inhomogeneous coordinates x and y on P_2, the curve S can be given by

$$x + 1 = 0, \quad y + 1 = 0,$$
$$xy + 1 = 0 \quad \text{and} \quad x^2 + y^2 - 2 = 0$$

Let us call this orbifold $X(\sqrt{2})$. Here is a picture of the singular locus S.

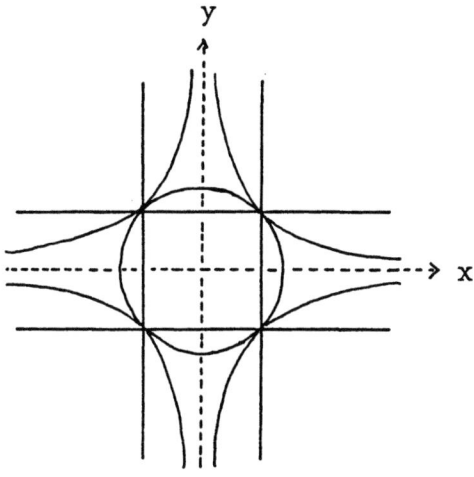

Orbifold $X(\sqrt{2})$

This is an orbifold obtained as the quotient of $H \times H$ by a Hilbert modular group Γ defined as follows. Let O be the ring of integers in the number field $\mathbb{Q}(\sqrt{2})$. The group $SL(2,O)$ acts as a discrete transformation group on $H \times H$ by

$$(z_1, z_2) \mapsto (g_1 z_1, g_2 z_2) \qquad g \in SL(2,O)$$

where $g \to g_1$ and $g \to g_2$ are two embeddings of $SL(2,O)$ into $SL(2,\mathbb{R})$. Denote by $\Gamma(2)$ the principal congruence subgroup of $SL(2,O)$ associated with the ideal (2) of O:

$$\Gamma(2) = \{ g \in SL(2,O) \mid g \equiv \begin{pmatrix} 1 & 0 \\ 0 & 1 \end{pmatrix} \mod (2) \}$$

Now let $\Gamma(2) \subset \Gamma' \subset SL(2,O)$ be the group such that $\Gamma'/\Gamma(2)$ is the center of $SL(2,O)/\Gamma(2)$. We note that $[\Gamma' : \Gamma(2)] = 2$ and $SL(2,O)/\Gamma' \cong S_4$ (symmetric group). Finally let $\Gamma = \langle \Gamma', \tau \rangle$ where τ is the involution $(z_1, z_2) \mapsto (z_2, z_1)$. Hirzebruch shows that the quotient space $H \times H / \Gamma$ is isomorphic to $\mathbb{P}_2$ minus six points; that the ramification locus of the natural projection $H \times H \to \mathbb{P}_2$ is exactly $S \subset \mathbb{P}_2$; and that the above six points are the six multiple points of S.

§ 10.3 Line Arrangements in P_2 and Orbifolds Covered by B_2

Following Hirzebruch and Höfer ([Hir 1,3],[Höf] and [BHH]) we shall find line arrangements A in P_2 such that there are orbifolds (X,S,b) covered by the ball and that $X-S = P_2-A$. Let A be an arrangement of k distinct lines $L_1,\ldots,L_k$ in the projective plane P_2. For a point p in the plane, we let r_p be the number of lines in the arrangement passing through p and t_r (for $t \geq 2$) be the number of points p with $r_p = r$. Then we have

$$k(k-1) = \Sigma_{r \geq 2} \, t_r \, r(r-1).$$

Given an arrangement, we blow up the nondouble multiple points p_j in the plane (i.e. those with $r_{p_j} \geq 3$) to get an algebraic surface X, in which we have a configuration of curves : the strict transforms of the lines of the arrangement (which we also call $L_1,\ldots,L_k$ and which are smooth irreducible curves on X) and the curves $E_1,\ldots,E_s$ obtained by blowing up the points p_j where $1 \leq j \leq s$. (Here s is the number of nondouble multiple points, in other words $s = \Sigma_{r \geq 3} \, t_r$.) We now associate weights $n_1,\ldots,n_k \in \{2,3,\ldots,\infty\}$ to the "old" lines $L_1,\ldots,L_k$ and weights $m_1,\ldots,m_s \in \{1,2,\ldots,\infty\}$ to the "new" lines $E_1,\ldots,E_s$. This defines an orbifold (X,S,b) with

$$S = \bigcup_i L_i \cup \bigcup_j \{E_j \mid m_j \geq 2\}$$

Let σ_i be the number (without counting multiplicities) of nondouble multiple points lying on L_i. Consider the $k \times k$ symmetric matrix A defined by

$$A_{ij} = \begin{cases} 3\sigma_i - 4 & i = j \\ 2 & i \neq j, \; p = L_i \cap L_j \text{ with } r_p = 2 \\ -1 & i \neq j, \; p = L_i \cap L_j \text{ with } r_p \geq 3 \end{cases}$$

Associate real variables x_i to the k lines and x be the column vector $(x_1,\ldots,x_k)$. Associate real variables y_j to the s nondouble multiple points p_j and let $y = (y_1,\ldots y_s)$. For the nondouble multiple point p_j, we consider the linear form

$$P_j(x,y) = 2y_j + \Sigma\, x_i \quad (p_j \in L_i).$$

Following Höfer and Hirzebruch we now define the quadratic form in real variables x_i and y_j:

$$\text{Prop}(x,y) = 4^{-1}({}^t x A x + \Sigma_{j=1}^s P_j(x,y)^2).$$

Observe that this quadratic form depends only on the (unweighted) arrangement. We now have the following useful formula:

Höfer's Formula:

$$3c_2(X,b) - c_1(X,b)^2$$
$$= \text{Prop}\,(1 - \tfrac{1}{n_1},\ldots,\, 1 - \tfrac{1}{n_k}, -1 - \tfrac{1}{m_1},\ldots,-1 - \tfrac{1}{m_s})$$

This formula explains the notation $\text{Prop}(x,y)$ for the quadratic form which gives the deviation from the proportionality $3c_2 = c_1^2$.

<u>Note</u>: By using the construction of the orbifold (X,S,b), we know all the quantities involved in the definition of $c_2(X,b)$

and $c_1^2(X,b)$. Thus it is possible to derive the formula by computation using only the (so-called) adjunction formula. However, we do not give this proof here. I think it is better to consult the book [BHH] or Hofer's original paper [Höf] than to give a purely computational proof; [BHH] and [Höf] give a good explanation of the geometrical meaning of this formula.

The only known arrangements, such that for a suitable choice of n_i's and m_j's we have $3c_2(X,b) = c_1^2(X,b)$, are as follows. (See also Example 9 at the end of this section.) The appropriate values of the n_i's and m_j's will be given later.

Examples (1) The complete quadrilateral (cf. § 10.2, § 10.5):

$$k = 6, \quad t_2 = 3, \quad t_3 = 4, \quad t_r = 0 \text{ otherwise}$$

In homogeneous coordinates z_0, z_1, z_2, the 6 lines can be given by the equation

$$z_0 z_1 z_2 (z_0 - z_1)(z_1 - z_2)(z_2 - z_0) = 0$$

(2) The arrangements $A_3^0(m)$ ($m \geq 3$) (cf. § 11.4, § 12.3):

$$k = 3m, \quad t_2 = 0, \quad t_3 = m^2, \quad t_m = 3, \quad t_r = 0 \text{ otherwise}$$

The 3m lines can be given by the equation

$$(z_0^m - z_1^m)(z_1^m - z_2^m)(z_2^m - z_0^m) = 0$$

(3) The arrangement $A_3^3(m)$ $(m \geq 3)$ (cf. § 11.4, § 12.3):

$$k = 3m+3, \; t_2 = 3m, \; t_3 = m^2, \; t_{m+2} = 3, \; t_r = 0 \text{ otherwise}$$

The $3m + 3$ lines can be given by the equation

$$z_0 z_1 z_2 (z_0^m - z_1^m)(z_1^m - z_2^m)(z_2^m - z_0^m) = 0$$

(4) The icosahedral arrangement (cf. § 11.4, § 12.4):

$$k = 15, \; t_2 = 15, \; t_3 = 10, \; t_5 = 6, \; t_r = 0 \text{ otherwise}$$

(5) The Hesse arrangement (cf. § 11.4, § 12.5):

$$k = 12, \; t_2 = 12, \; t_4 = 9, \quad t_r = 0 \text{ otherwise}$$

The 12 lines can be given by the equations

$$z_j = 0 \qquad (j = 0,1,2)$$
$$z_0 + \omega^a z_1 + \omega^b z_2 = 0 \qquad (a,b = 0,1,2)$$

where ω is a cube root of unity.

(6) The extended Hesse arrangement (cf. § 11.4, § 12.5):

$$k = 21, \; t_2 = 36, \; t_4 = 9, \; t_5 = 12, \; t_r = 0 \text{ otherwise}$$

The 21 lines can be given by adding the 9 extra lines $A_3^0(3)$ to the Hesse arrangement.

(7) The Klein arrangement (cf. § 11.4, § 12.4):

$$k = 21, \; t_3 = 28, \; t_4 = 21, \quad t_r = 0 \text{ otherwise}$$

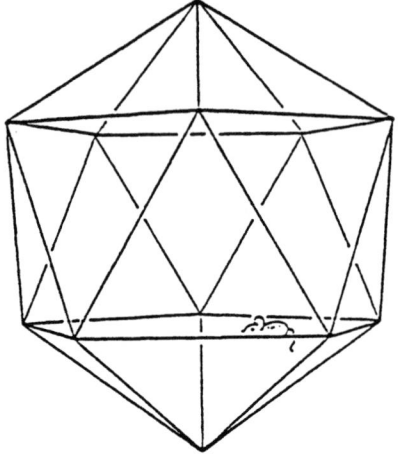

The icosahedron

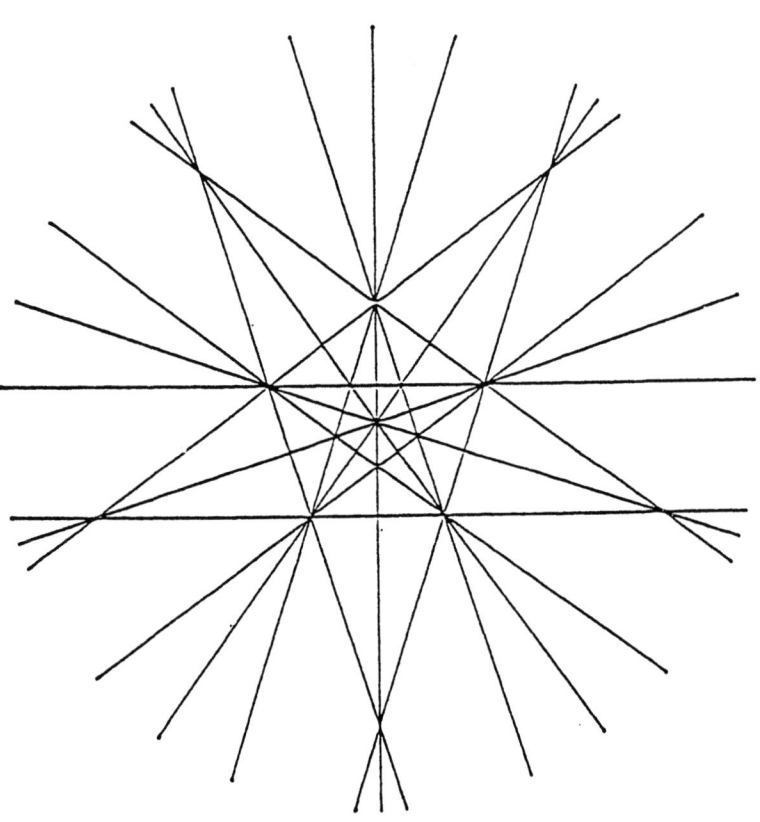

The icosahedral arrangement

(8) <u>The Valentiner arrangement</u> (cf. § 11.4, § 12.4):

$$k = 45, \ t_3 = 120, \ t_4 = 45, \ t_5 = 36, \ t_r = 0 \text{ otherwise}$$

<u>Remarks</u> 1) Examples (1) and (5) are real arrangements, that is, the L_j's are defined over the reals.

2) Let τ_j be the number of multiple points on L_j. For every example above, one can check that

$$3\tau_j = k + 3 \qquad (j = 1, \ldots, k).$$

We can show, for every example above, that the quadratic form A is positive semi-definite and hence is Prop(x,y). Hence the homogeneous linear equations in the k + s real variables $x_1, \ldots, x_k, \ y_1, \ldots, y_s$:

(10.1) $\qquad Ax = 0, \qquad P_j(x,y) = 0 \qquad (j = 1, \ldots, s)$

define the nullspace of Prop(x,y) whose dimension equals the corank of A. Therefore the orbifold (X,S,b) satisfies $3c_2(X,b) = c_1(X,b)^2$ if and only if

(10.2) $\qquad x_i = 1 - 1/n_i \qquad\qquad (2 \leq n_i \leq \infty)$
$\qquad\qquad\quad y_j = -1 - 1/m_j \qquad\quad (1 \leq m_j \leq \infty)$

is a solution of (10.1).

<u>Note</u>: By using the equations $P_j(x,y) = 0$, the n_j's determine the m_i's. In general we do not get the same values as in § 10.1 in Kato's existence theorem.

For the complete quadrilateral, corank A = 4, there are finitely many solutions of (10.1) and (10.2). We come back to this later in § 10.5.

For the arrangements $A_3^0(m)$ and $A_3^3(m)$, let us consider the map $\mathbb{P}_2 \to \mathbb{P}_2$ defined by

$$\psi : (z_0, z_1, z_2) \mapsto (w_0, w_1, w_2) = (z_0^m, z_1^m, z_2^m)$$

i.e. the quotient map of the cyclic group G acting by

$$(z_0, z_1, z_2) \mapsto (\theta z_0, \theta z_1, \theta z_2) \qquad (\theta^m = 1)$$

All the solutions (n_i, m_j) of (10.1) and (10.2) associated to the arrangements $A_3^0(m)$ and $A_3^3(m)$ are G-invariant i.e. for each i,j ($0 \le i,j \le 2$, $i \ne j$) the weights attached to the m lines $z_i = \theta z_j$ ($\theta^m = 1$) are the same. Thus the corresponding orbifolds are obtained by lifting (via the map ψ) those associated to the quadrilateral arrangement. (cf. § 12.3)

For the extended-Hesse arrangement, corank A = 2. The x_i's are constant for the lines of the Hesse arrangement and also constant for the additional 9 lines. There are exactly seven solutions of (10.1) and (10.2). The weights n_i are

Hesse lines	additional lines
2	∞
2	2
2	3
3	9
4	2
4	6
∞	3

For the other arrangements namely for examples (4),(5),(7) and (8), corank A = 1 and the x_i's are independent of the i's. This follows from Remark 2 above and the following computation:

(sum of the entries of i-th line of the matrix A)
= -k + 3 #{nondouble multiple points} - 3
+ 3 #{double multiple points}
= -(k + 3 - 3τ_i)

The possible weights $n = n_i$ (i = 1,...,k) are tabulated below. If we permit negative m_j's in (10.2), there are other solutions of (10.1) and (10.2) which are also tabulated below, and indicated by the symbol *. The meaning of these three extra solutions is explained after the table.

Arrangement	possible values of n
Hesse	2, 3, 4
Icosahedral	2*, 5, ∞
Klein	2*, 3, 4
Valentiner	2*

Until now, in this paragraph, we have been studying orbifolds (X,S,b) whose base spaces X are obtained by blowing up <u>all</u> the nondouble multiple points of line arrangements A. But this is not always necessary to obtain an orbifold. We shall study in § 11.2 admissible singularities of S. Here we quote some results.

" Consider t (≥ 3) lines $S_1,\ldots,S_t$ meeting at a point in some domain $D \subset \mathbb{C}^2$. If one associates weight b_j to S_j (j = 1,...,t) then the orbifold (D,S,b) is uniformizable if and only if $\Sigma (1 - \frac{1}{b_j}) \geq 2$. (This implies t = 3 or 4.)"

We now consider orbifolds whose base spaces are obtained by blowing up only some of the nondouble multiple points of the arrangements. We make a slight modification of the definition of c_1^2 and c_2 for (X,S,b) in order S to admit nondouble multiple points; then R. Kobayashi's theorem (§ 10.2) still holds and by repeating the arguments of this section we find the following result:

" For a line arrangement A, assume that (10.1) and (10.2) have an integral solution n_i, m_j ($2 \leq n_i \leq \infty$, $m_j \in \mathbb{Z} \cup \{\infty\}$). When constructing the orbifold according to the procedure at the begining of this section, blow up only those nondouble multiple points p_j such that $m_j \geq 1$ to get the surface X. Associate weights n_i to L_i and weights m_j (≥ 2) to the "new" curves E_j. Then the orbifold (X,S,b) thus obtained is uniformized by P_2 or B_2."

Remark 3): All the examples of orbifolds obtained in this section are uniformized by B_2. This fact is checked, for each orbifold $\underline{X}$ = (X,S,b), by showing that the universal uniformization M can not be P_2. For each $\underline{X}$, we can find m_j which is either infinity or an integer (≥ 2). If $m_j = \infty$, then the projection π: M → X is an infinite covering; so M can not be P_2. If $m_j \geq 2$, assume M = P_2 and consider a loop going around once the curve E_j in X. The corresponding transformation of M fixes a line $C \subset M$ and the restriction on C of π : C → E_j is a finite covering of P_1 by P_1. Applying Hurwitz' formula: $-2 + 2s = \Sigma \frac{s}{n_i} (n_j - 1)$ ($p_j \in L_j$), where s is the number of sheets of the covering, we have $\Sigma(1 - \frac{1}{n_i}) < 2$ ($p_i \in L_j$). On the other hand, m_j is determined by $0 = P_j = 2(-1 - \frac{1}{m_j}) + \Sigma(1 - \frac{1}{n_i})$ ($p_i \in L_j$). This is a contradiction.

In his theses [Hun], B. Hunt studied an N-dimensional version of the argument of this section. Among others he discovers a 3-dimensional orbifold, uniformized by the 3-ball B_3, attached to an arragement of planes in P_3. This plane arrangement is called the F_4-arrangement and consists of 24 planes which are the mirrors of the 4-dimensional reflection group F_4 (see the end of Chapter 11). Restricting to a plane in the arrangement, it gives a 2-dimensional orbifold, uniformized by B_2, attached to a line arrangement in P_2 which is also called the F_4-arrangement.

Example (9) The F_4 arrangement (cf. § 11.4 and [Hun]):

$$k = 13, \quad t_2 = 12, \quad t_3 = 4, \quad t_4 = 9, \quad t_r = 0 \quad \text{otherwise}$$

Since this arrangement is defined over the reals, we illustrate it in the next page in which lines are labelled by a, b and c; and nondouble multiple points are labelled by A, B and C. The orbifold is obtained by blowing up the nine 4-fold points (marked B and C) and choosing weights as follows

lines	weights n_i	points	weights m_j
a	∞	A	-4
b	2	B	2
c	4	C	2

Then the n_i's and m_j's satisfy (10.1) and (10.2). Therefore the orbifold is uniformized by B_2. Höfer kindly informed me that there are only three other orbifolds covered by a ball over by this arrangement. Their weights are given respectively as follows

lines	weights n_i	points	weights m_j
a	6, -6, -3	A	-4, 4, ∞
b	2, 6, 3	B	6, 2, 1
c	2, 2, ∞	C	3, 1, 1

134

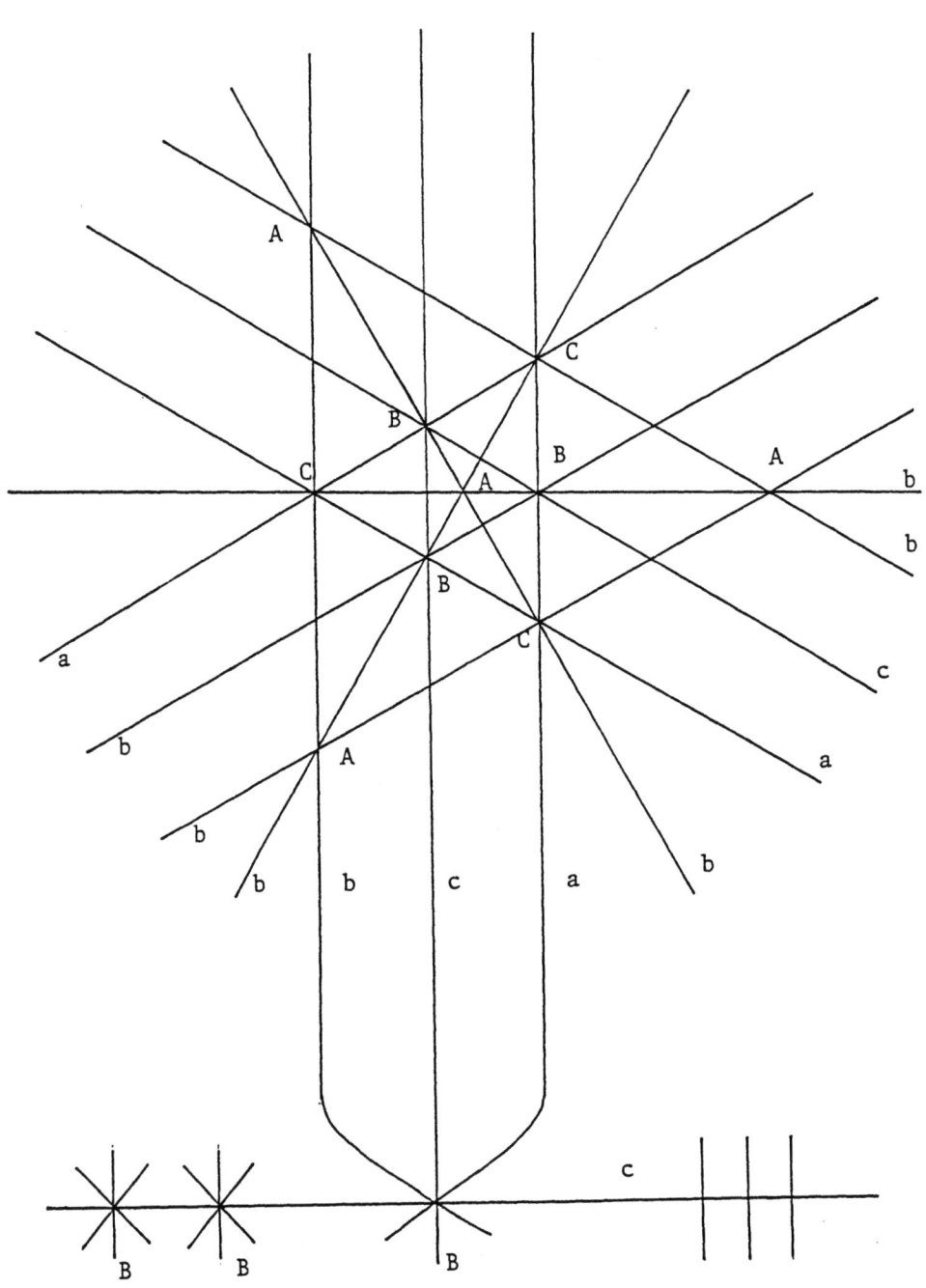

§ 10.4 Uniformizing Differential Equations

Let (X,S,b) be an uniformizable orbifold, M be the universal uniformization with the covering map

$$\pi : M \to X$$

and z be the developing map (i.e. the inverse of π). Fix a coordinate system (z^1, z^2) on M and cover X by coordinate neighbourhoods U_α with local coordinate systems $x_\alpha = (x^1, x^2)$.

Let us consider firstly the cases $M = \mathbf{P}_2$, $\mathbb{C}^2$ or $\mathbf{B}_2$. Since we have

$$\text{Aut}(M) \subset \text{PGL}(3,\mathbb{C}),$$

we can define single-valued meromorphic functions $p_{\alpha ij}^k$ by

$$p_{\alpha ij}^k(x_\alpha) = S_{ij}^k(z; x_\alpha) \qquad i,j,k = 1,2.$$

The projective equation defined by the system $\{E_\alpha\}$ of integrable (i.e. the rank is 3) differential equations E_α on U_α with coefficients $p_{\alpha ij}^k = p_{\alpha ij}^k(x_\alpha)$:

$$\frac{\partial^2 u}{\partial x_\alpha^i \partial x_\alpha^j} = \Sigma_{k=1}^2 p_{\alpha ij}^k \frac{\partial u}{\partial x_\alpha^k} + p_{\alpha ij}^0 u \qquad i,j = 1,2$$

is called <u>the uniformizing differential equation of the orbifold</u> (X,S,b). These equations have ramifying singularities of exponent $1/b_j$ along S_j. Since they satisfy

$$p_{\alpha ij}^k = p_{\alpha ji}^k \qquad \text{and the connection formula:}$$

$$p^k_{\alpha ij} - p^c_{\beta ab}\frac{\partial x^a_\beta}{\partial x^i_\alpha}\frac{\partial x^b_\beta}{\partial x^j_\alpha}\frac{\partial x^k_\alpha}{\partial x^c_\beta} = S^k_{ij}(x_\beta;x_\alpha),$$

the system $\{p^k_{\alpha ij}(x_\alpha)\}$ defines a flat meromorphic connection. Furthermore, a finite uniformization $Y \to X$ always exists (cf. Selberg's theorem § 5.1) and the induced system on Y then defines a holomorphic flat projective connection on Y.

If $X-S = P_2-A$, then the uniformizing equation is also defined on P_2 and has regular singularities along A. Notice that since P_2 can be covered by projectively related coordinates, the following meromorphic <u>tensor</u> field

$$p^k_{\alpha ij}\, dx^i_\alpha \otimes dx^j_\alpha \otimes \frac{\partial}{\partial x^k_\alpha}$$

determines the uniformizing differential equation (see § 12.1).

Consider next the case $M = H \times H$ or $P_1 \times P_1$. Since $P_1 \times P_1$ can be regarded as a non-degenerate quadric surface Q in P_3, the group $\text{Aut}(P_1 \times P_1)$ is the subgroup $PO(Q)$ of $PGL(4,\mathbb{C})$ which preserves Q. Hence we have

$$\text{Aut}(M) \subset \text{Aut}(P_1 \times P_1) \cong PO(Q) \subset PGL(4,\mathbb{C}).$$

The uniformizing equations are therefore not of the form E but of the form

$$\text{(EQ)} \quad \begin{aligned} u_{11} &= a^0 u + a^1 u_1 + a^2 u_2 + a^{12} u_{12} \\ u_{22} &= b^0 u + b^1 u_1 + b^2 u_2 + b^{12} u_{12} \end{aligned}$$

of rank 4 such that 4 linearly independent solutions satisfy a quadratic relation (cf. § 6.2.). In [S-Y], a systematic study of such equations is made. An explicit form for the uniformizing equation for the orbifold $X(\sqrt{2})$ (introduced at the end of § 10.2.) is also obtained; which we reproduce the result here.

Theorem: The coefficients of the uniformizing equation of the form (EQ) for the orbifold $X(\sqrt{2})$ are given as follows:

$$a^0 = -\frac{2-y^2-x^2y^2}{xy(1-x^2)}$$

$$a^1 = -\frac{3}{2}\frac{\partial}{\partial x}\log\frac{(1-x^2y^2)(2-x^2-y^2)}{1-x^2}$$

$$+ \frac{a^0}{2}\frac{\partial}{\partial y}\log\frac{(1-x^2y^2)^2(2-x^2-y^2)^2}{(1-y^2)^2(2-y^2-x^2y^2)}$$

$$a^2 = \frac{a^0}{2}\frac{\partial}{\partial x}\log\frac{(2-y^2-x^2y^2)(1-x^2y^2)(2-x^2-y^2)}{(1-x^2)^2}$$

$$a^{12} = \frac{-2(x^2-y^2)}{(1-x^2)^2(1-y^2)}$$

The coefficients b^0, b^1, b^2 and b^{12} are obtained from a^0, a^2, a^1 and a^{12}, respectively, by exchanging x and y.

§ 10.5 Gauss-Schwarz Theory for Appell's F_1

We study the uniformizing differential equations of orbifolds (X,S,b) such that $X-S = \mathbf{P}_2 - A$ where A is the complete quadrilateral defined by

$$x^0 x^1 x^2 (x^0 - x^1)(x^1 - x^2)(x^2 - x^0) = 0.$$

The differential equations are Fuchsian with ramifying singularities along A and they turn out to be Appell's system F_1 as we see in § 12.3. For the orbifolds related to the other line arrangements that appeared in § 10.3, we shall find their uniformizing equations in Chapter 12, in preparation we look at unitary reflection groups in the next chapter.

Let us study the singularities of the system $F_1(a,b,b',c)$

$$x^1(1 - x^1)u_{11} + x^2(1 - x^1)u_{12} + (c - (a+b+1)x^1)u_1$$
$$- bx^2 u_2 - ab = 0$$
$$x^2(1 - x^2)u_{22} + x^1(1 - x^2)u_{21} + (c - (a+b'+1)x^2)u_2$$
$$- bx^1 u_1 - ab' = 0$$
$$(x^1 - x^2)u_{12} - b'u_1 + bu_2 = 0$$

with complex parameters a,b,b',c, and independent variables x^1, x^2. Transform the system into the canonical form by the formula in the proposition of § 8.3. The coefficients p_{ij}^k in canonical form can be calculated from p_{11}^1, p_{11}^2, p_{22}^1 and p_{22}^2 which are given by

$$p_{11}^1 = \frac{1}{3} \frac{x^2(1 - x^2)}{f} \{ (c-b')x^2 + (2b-c)x^1$$
$$+ (b'-(a+b+1))x^1 x^2 + (a-b+1)(x^1)^2 \}$$
$$p_{11}^2 = \frac{(x^2(1 - x^2))^2}{f} b$$
$$p_{22}^2 = -\frac{1}{3} \frac{x^1(1 - x^1)}{f} \{ (c-b)x^1 + (2b'-c)x^2$$
$$+ (b-(a+b'+1))x^2 x^1 + (a-b'+1)(x^2)^2 \}$$
$$p_{22}^1 = -\frac{(x^1(1 - x^1))^2}{f} b'$$

where $f = x^1 x^2 (1-x^1)(1-x^2)(x^1-x^2)$. Since we have

$$x^1 \; p_{11}^1 \big|_{x^1 = 0} = \frac{1}{3}(-1)(c-b'),$$

the equation has ramifying singularities with exponent $b'-c+1$ along the line $x^1 = 0$.

Next we study the symmetricity of the system. Transform the Euler integral representation (§ 6.4)

$$\int t^{a-1}(1-tx^1)^{-b}(1-tx^2)^{-b'}(1-t)^{c-a-1} \, dt$$

by putting $t = 1/s$, then we have

$$\int s^{b+b'-c}(s-x^1)^{-b}(s-x^2)^{-b'}(s-1)^{c-a-1} \, ds.$$

Put

$$\lambda_0 = b+b'-c+1, \quad \lambda_1 = 1-b, \quad \lambda_2 = 1-b'$$
$$\lambda_3 = c-a, \quad \lambda_4 = a,$$

or equivalently,

$$a = \lambda_4, \quad b = 1-\lambda_1, \quad b' = 1-\lambda_2, \quad c = \lambda_3 + \lambda_4,$$
$$\lambda_0 + \lambda_1 + \lambda_2 + \lambda_3 + \lambda_4 = 3$$

and denote the system $F_1(a,b,b',c)$ by $F_1(\lambda) = F_1(\lambda_0,\ldots,\lambda_4)$. Then the Euler integral above is an integral of a (multivalued) 1-form which has singularities at $s = 0$, x^1, x^2, 1 and ∞ with exponents λ_0-1, λ_1-1, λ_2-1, λ_3-1 and λ_4-1, respectively.

Accordingly, let us introduce five coordinates

$$x = (x^0, x^1, x^2, x^3, x^4)$$

and consider the space

$$X = \{x \in (\mathbf{P}_1)^5 | \text{ no 3 coordinates are equal }\} / \text{Aut}(\mathbf{P}_1)$$

where $\text{Aut}(\mathbf{P}_1)$ acts diagonally on $(\mathbf{P}_1)^5$. i.e.

$$\sigma(x^0,\ldots,x^4) = (\sigma x^0,\ldots,\sigma x^4) \qquad \sigma \in \text{Aut}(\mathbf{P}_1).$$

Define the subspaces $S(ij)$ of X by the equation $x^i = x^j$ ($0 \leq i \neq j \leq 4$). Then we have

<u>Lemma</u>: (i) The space X equipped with the quotient topology has the natural structure of a projective algebraic surface. (ii) The symmetric group S_5 acts biholomorphically on X by permuting the five coordinates. (iii) X is biholomorphic to the surface obtained by blowing up four points of $\mathbf{P}_2$ in general position (i.e. no three points lie on a line). (iv) Each $S(ij)$ is a nonsingular rational curve of self-intersection number -1. (v) The intersection pattern of ten curves $S(ij)$ is given as follows

$$S(ij) \cap S(kl) = \begin{cases} \phi & \text{if } \{i,j\} \cap \{k,l\} \neq \phi \\ \text{a point} & \text{if } \{i,j\} \cap \{k,l\} = \phi \end{cases}$$

<u>Proof</u>: Let the 5-tuple (i,j,k,l,m) be a permutation of $0, 1, 2, 3$ and 4. Let $U(ij)$ be the part of X consisting of $x = (x^0,\ldots,x^4)$ such that x^k, x^l and x^m are distinct. Since we can always find three distinct coordinates in $(x^0,\ldots,x^4) \in X$, X is covered by the $U(ij)$'s. In each $U(ij)$, by normalizing the three distinct coordinates as $x^k = 0$, $x^l = 1$, $x^m = \infty$, we see that it is homeomorphic to the surface $\mathbf{P}_1 \times \mathbf{P}_1 - \{(0,0), (1,1), (\infty,\infty)\}$ parameterized by (x^i, x^j). This gives a natural nonsingular analytic structure on X. The set $S(ij)$ has the structure of a projective line because after normalizing as $x^i = x^j = \infty$ and $x^k = 0$, it is given by $\{(x^l, x^m) \in \mathbb{C}^2 - \{(0,0)\}\}/\mathbb{C}^*$. Since we have $X - U(ij)$

= $S(kl) \cup S(lm) \cup S(mk)$, we see that X is isomorphic to the surface obtained by blowing up three points $(0,0)$, $(1,1)$ and (∞,∞) of $P_1 \times P_1$. Blowing up $P_1 \times P_1$ at these three points gives the same space as blowing up P_2 at four points in general position. The other claims are now obvious.

Let us blow down the four disjoint curves $S(m4)$ ($0 \leq m \leq 3$) to get P_2 in which the images of the remaining six curves $S(ij)$ ($0 \leq i,j \leq 3$) (which we also denote by the same symbols) form the complete quadrilateral line arrangement A. If we denote the image of $S(m4)$ by $S(ijk)$ ($\{i,j,k,m\} = \{0,1,2,3\}$) then the map $X \to P_2$ is illustrated as follows.

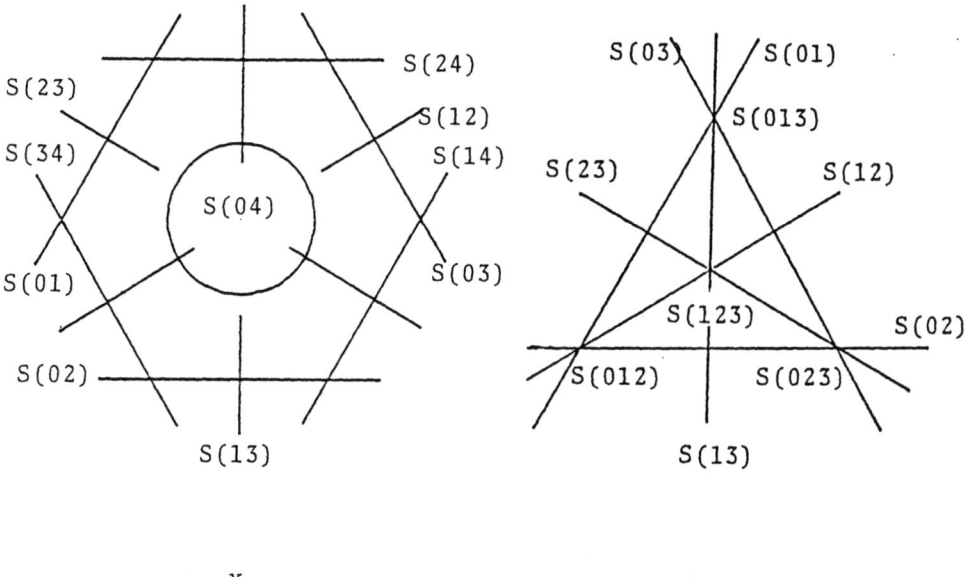

$$X \longrightarrow P_2$$

The system F_1 is now defined on X. Obviously the symmetric group S_5 acts transitively on the 10 curves $S(ij)$. Since the equation $F_1(\lambda)$ has solutions represented by integrals of the form

$$\int (s-x^0)^{\lambda_0-1}(s-x^1)^{\lambda_1-1}\cdots(s-x^4)^{\lambda_4-1}\,ds$$

the family of systems $\{F_1(\lambda_0,\ldots,\lambda_4)\}$ ($\sum_{j=0}^4 \lambda_j = 3$) is invariant under S_5. We have already calculated the exponent along $S(01)$ to be $b' - c + 1 = \lambda_0 + \lambda_1 - 1$. Hence we have

Lemma: The system $F_1(\lambda)$ on X has ramifying singularities of exponents $\lambda_i + \lambda_j - 1$ along $S(ij)$ ($0 \leq i, j \leq 4$; $i \neq j$).

Let (X,S,b) be an orbifold with weights $b(ij)$ on $S(ij)$. If it is uniformized by the ball, then the uniformizing differential equation must be the system $F_1(\lambda)$ for some λ. Thus there are numbers $\lambda_0,\ldots,\lambda_4$ ($\sum \lambda_j = 3$) such that

$$b(ij) = (\lambda_i + \lambda_j - 1)^{-1}.$$

Conversely if there are numbers $\lambda_0,\ldots,\lambda_4$ ($\sum \lambda_j = 3$) such that $(\lambda_i + \lambda_j - 1)^{-1} \in \{2, 3, \ldots, \infty\}$ then defining the weights $b(ij)$ by the above formula, the orbifold (X,S,b) thus defined is covered by $\mathbf{P}_2$ or $\mathbf{B}_2$. Indeed since we have

$$2b(01)^{-1} + b(23)^{-1} + b(24)^{-1} + b(34)^{-1} = 1$$

and all its permutations, Höfer's formula tells us $3c_2(X,b) = c_1(X,b)^2$. One can check that, up to permutation, there are 12 sets of rational numbers $\{\lambda_0,\ldots,\lambda_4\}$ satisfying these conditions. Appell's equation F_1 uniformizes not only these orbifolds but also those for which some of the $b(ij) = (\lambda_i + \lambda_j - 1)^{-1}$ are negative integers. Therefore we have proved the following theorem.

Theorem (Picard-Terada-Mostow-Deligne): Let $\lambda_0,\ldots,\lambda_4$ be reals with $0 < \lambda_j < 1$ and $\Sigma \lambda_j = 3$ such that the following condition holds: " for all i,j ($0 \leq i \neq j \leq 4$) we have { $\lambda_i + \lambda_j - 1)^{-1} \varepsilon \mathbb{Z} \cup \{\infty\}$." Define $X(\lambda)$ to be the surface obtained from X by blowing down the curves $S(ij)$ with $\lambda_i + \lambda_j - 1 < 0$. Then the orbifold on $X(\lambda)$ associating the weight $b(ij) = (\lambda_i + \lambda_j - 1)^{-1}$ to the curve $S(ij)$ with $\lambda_i + \lambda_j - 1 \geq 0$ is uniformized by the ball. The uniformizing equation is Appell's system $F_1(\lambda)$.

We refer this theorem as the **PTMD-theorem**. Up to permutation there are 27 sets of λ_j's satisfying the hypotheses of this theorem. They are listed below

d = lowest common denominator of $\lambda_0,\ldots,\lambda_4$

∞ means some $b(ij)$ is infinity

d		$d\lambda_j$				
3	1	2	2	2	2	∞
4	2	2	2	3	3	∞
4	1	2	3	3	3	∞
5	3	3	3	3	3	
6	3	3	4	4	4	∞
6	3	3	3	4	5	∞
6	2	3	4	4	5	∞
6	1	4	4	4	5	∞
8	4	5	5	5	5	
8	3	3	6	6	6	
8	2	5	5	5	7	
9	5	5	5	5	7	
10	3	6	6	6	9	
12	7	7	7	7	8	
12	6	7	7	8	8	
12	6	7	7	7	9	

d	$d\lambda_j$					
12	5	7	8	8	8	∞
12	5	6	7	9	9	∞
12	5	5	8	8	10	
12	4	7	7	9	9	
12	4	7	7	7	11	
12	4	5	9	9	9	
12	2	7	9	9	9	
15	7	9	9	9	11	
18	7	10	10	10	17	
20	6	9	15	15	15	
24	10	15	15	15	17	

Historical remark: E.Picard treated the case $\lambda_0 = \lambda_1 = \lambda_2 = \lambda_3 = 2/3$, $\lambda_4 = 1/3$. He studied the local and global behaviour of the projective solution z of Appell's equation $F_1(\lambda)$ but he could not prove rigorously that it gives the developing map of the orbifold $X(\lambda)$ (ie. the image of the map z is the ball). The Gauss-Schwarz theory for HGDE in single variable was based on (Schwarz's) reflection principle or the uniformization theorem (of Riemann surfaces). In several variables, there are no theorems analogous to these two. Therefore there was no method at that time to prove the two dimensional analogue of the Gauss-Schwarz theory (PTMD-theorem). It was Terada([Ter 1]) who formulated the theorem and gave a proof for the compact case (ie. the case all weights S(i,j) are finite). Then Mostow-Deligne([D-M]) and Terada([Ter 2]) independently proved the theorem for all 27 cases. Terada's proof is function theoretic while Mostow-Deligne's is algebro-geometric. Although their proofs work only for the special arrangement, namely the complete quadrilateral, our differential geometric proof works for any arrangements; essentially, it depends on the existence of the Einstein-Kähler metric and the calculus of curvature integrals.

Remarks: 1) A higher dimensional analogue of the theorem is also established in [Ter 2] and [M-D].

2) (cf. § 11.2 Case $M = \mathbb{C}^2$) If $\lambda_0 = \lambda_1 = \lambda_2 = \lambda_3 = 3/4$ and $\lambda_4 = 0$ then

$$(\lambda_i + \lambda_j - 1)^{-1} = 2 \qquad 0 \leq i,j \leq 3$$
$$\lambda_m + \lambda_4 - 1 < 0 \qquad 0 \leq m \leq 3$$

In this case the surface $X(\lambda)$ is $\mathbf{P}_2$ itself and the orbifold on $X(\lambda)$ associating weight 2 to all 6 lines is uniformized by $\mathbb{C}^2$.

§ 10.6 Monodromy Representations of Appell's Equation

In this section we give a set of generators of the monodromy group of Appell's equation $F_1(\lambda)$. This is essentially obtained already by Picard. Here we follow the method used by Terada in [Ter 1]. Let the λ_j's, the x^j's, the $S(ij)$'s and X be as above. We assume in this section that the λ_j's are not integers. The equation has no singularities on

$$X_0 = X - \bigcup_{i,j} S(ij)$$
$$= \mathbf{P}_2 - A \quad \text{(where A is the complete quadrilateral)}$$

and its solutions are expressed as linear combinations of functions obtained by integrating

$$J = J(x,t) = \Pi_{j=0}^{4}(t - x^j)^{\lambda_j - 1}$$

along some paths in the t-plane. We are now going to construct holomorphic solutions in a contractible open neighbourhood U of a fixed point $a = (a^0,\ldots,a^4)$ in X_0. For any $x = (x^0,\ldots,x^4) \in U$ let $T(x)$ be a simply connected closed domain of P_1 (parameterized by t) whose boundary is a simple curve that depends continuously on x and passes through the points $x^0,\ldots,x^4 \in P_1$ in numerical order. Let $C_x(ij)$ ($0 \leq i \neq j \leq 4$) be a path in $T(x)$ starting x^i and ending at x^j; and put $C_x(j) = C_x(4j)$ ($0 \leq j \leq 3$). Since both U and $T(x)$ are simply connected, after fixing once and for all a branch of $J(x,t)$ on $\cup_{x \in U} \{x\} \times T(x)$, if we suppose that Re $\lambda_i > 0$ and Re $\lambda_j > 0$ we find that the integral

$$I(ij) = I(ij)(x) = \int_{C_x(ij)} J(x,t) dt$$

is a holomorphic function in U not depending on the chosen path $C_x(ij)$. Since the fuction $I(ij)$ is also meromorphic in λ_i and λ_j, by using analytic continuation, we extend $I(ij)$ for any $\lambda_i, \lambda_j \in \mathbb{C}$. This continuation will be denoted by the same symbol $I(ij)$. You can also use double contour loop (see § 1.4) around x^i and x^j. Since we have obvious relations

$$I(ij) + I(ji) = 0 \quad \text{and} \quad I(ij) + I(jk) + I(ki) = 0,$$

all the $I(ij)$'s can be expressed as linear combinations of the following four

$$I(j): = I(4j) = \int_{C_x(j)} J(x,t) dt \qquad (0 \leq j \leq 3)$$

In order to get a linear relation between the $I(j)$'s, consider a simple loop in P_1 disjoint from $T(x)$: such a loop is contractible, so the integral of J along the loop is zero. Now deform the loop in $P_1 - \{x^0,\ldots,x^4\}$ so that it is contained in T(x) except for small semi-circles around $x^0,\ldots,x^4$.

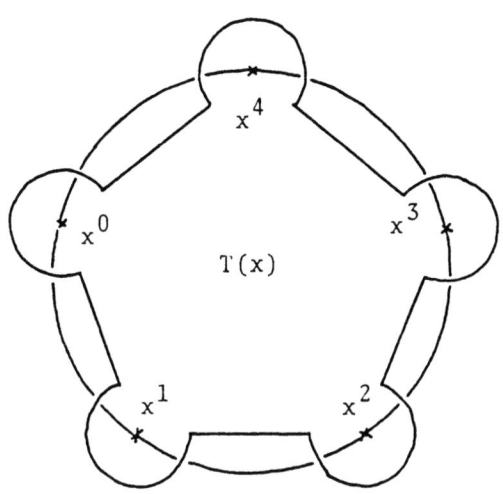

This deformation does not alter the integral of J(x,t) along the curve, which remains zero. Since the analytic continuation along a small positively directed circle around x^j changes the integrand J(x,t) into $e(\lambda_j)J(x,t)$, we have

$$I(40) + e(\lambda_0)I(01) + e(\lambda_0+\lambda_1)I(12) + e(\lambda_0+\lambda_1+\lambda_2)I(23)$$
$$+ e(\lambda_0+\lambda_1+\lambda_2+\lambda_3)I(34) = 0,$$

where $e(\lambda)$ stands for $\exp(2\pi i\lambda)$. This equality leads to a linear relation among the I(j)'s. We can check that any three of the I(j)'s are linearly independent.

<u>The fundamental group</u> $\Pi_1(X_0,a)$: Define a loop $\rho(ij)$ in X_0 $s \mapsto x(s) = (x^0,x^1,x^2,x^3,x^4)$ which starts at $x = a$ such that the coordinates x^k ($k \neq i$) are constant functions of s while x^i travels in T(a) up to a^j, goes once around a^j in the positive direction and travels back through T(a) to a^i. One can check that $\rho(ij) = \rho(ji)$.

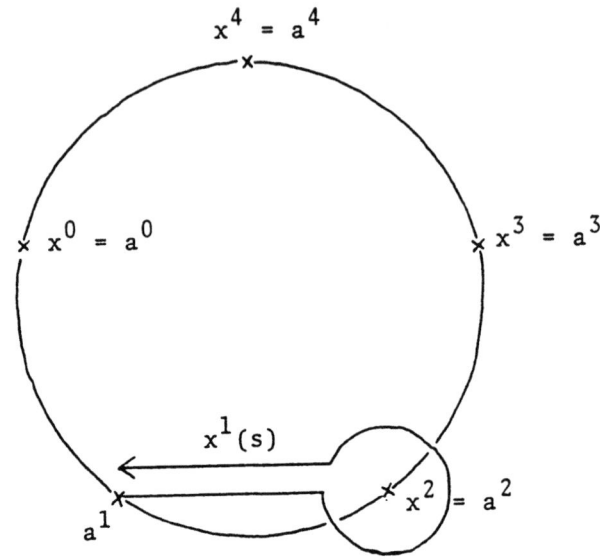

The fundamental group $\Pi_1(X_0,a)$ is generated by the $\rho(ij)$'s. As is easily seen, five of them are sufficient to generate it.

<u>Action of the $\rho(ij)$'s on the $I(k)$'s:</u> In view of the symmetricity with respect to the indices 0, 1, 2, 3 and 4, we need only consider the action of $\rho(12)$. Let us analytically continue the functions $I(j)$'s along the loop $\rho(12)$. The analytic continuation is found by deforming the paths of integration $C_x(j)$ in $P_1 - \{x^0,\ldots,x^4\}$ as $x = (x^0,\ldots,x^4) \in X_0$ travels along $\rho(12)$, namely as the point x^1 in P_1 follows a loop going once around the point x^2. To ensure that the moving point x^1 never meets any of the paths $C_x(j)$, not only $C_x(1)$ but also $C_x(2)$ is forced to make way for the passage of x^1. The paths $C_x(0)$ and $C_x(3)$ have nothing to do with the moving point x^1 so they can remain as they were. The following illustrations show these deformations.

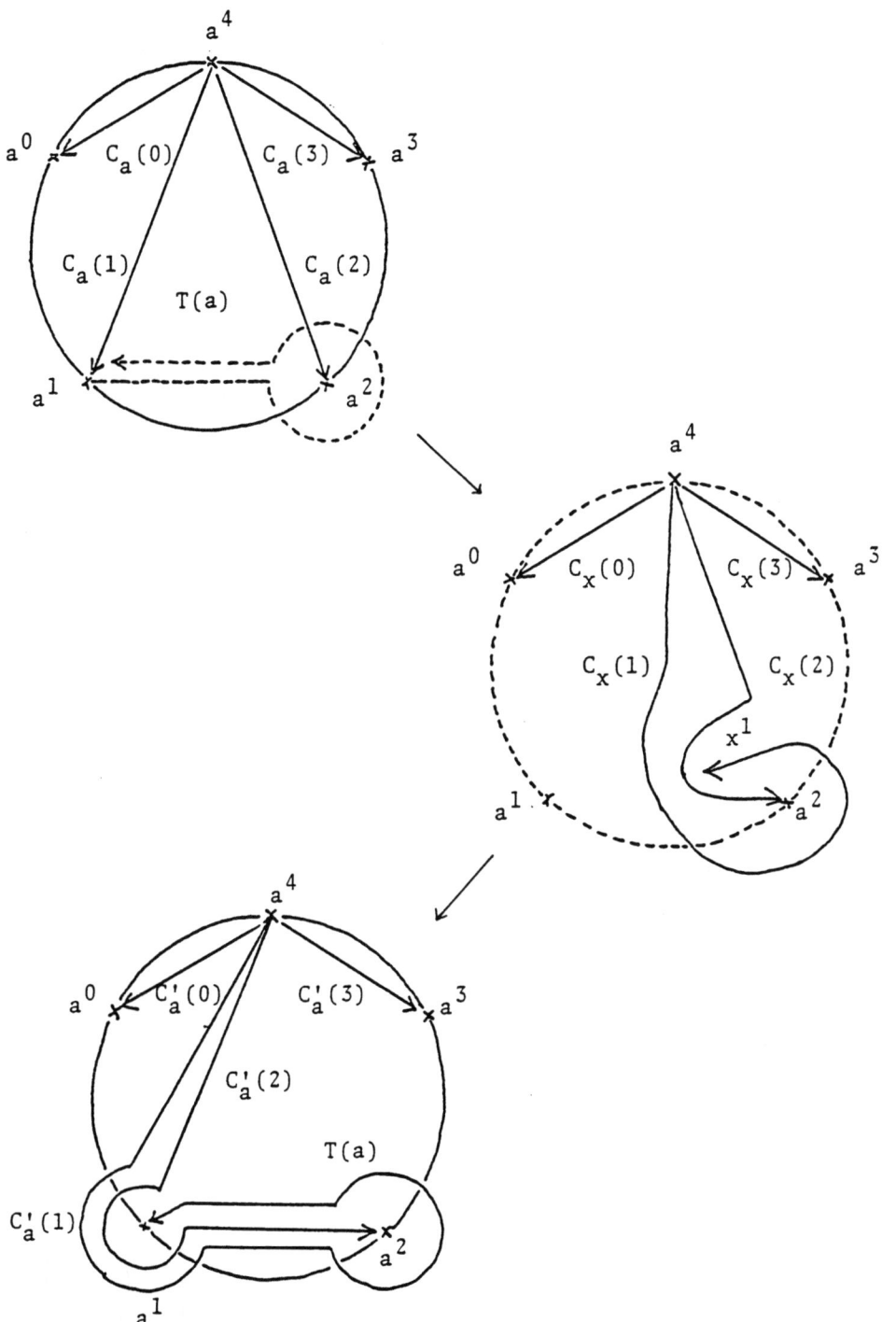

The last picture shows the result when the point x has ended its journey and the paths $C_a(j)$'s have been deformed into $C_a'(j)$'s. From these illustrations one can follow the changes in the $I(j)$'s:

$$\rho_*(12)I(0) = I(0), \qquad \rho_*(12)I(3) = I(3)$$
$$\rho_*(12)I(1) = I(1) + e(\lambda_1)I(12) + e(\lambda_1+\lambda_2)I(21)$$
$$\rho_*(12)I(2) = I(1) + e(\lambda_1)I(12)$$

where $\rho_*(ij)$ denotes the analytic continuation along $\rho(ij)$. Since we have $I(12) = I(1) - I(2)$, We know the action of $\rho(12)$ on the $I(j)$'s.

Generators of the monodromy group: Set for $1 \leq k \leq 3$

$$I_0(k) = (1-e(\lambda_k))^{-1}I(k) - (1-e(\lambda_0))^{-1}I(0)$$

and denote by $R(ij) \in GL(3,\mathbb{C})$ the circuit matrix of the three solutions $I_0(k)$ along $\rho(ij)$, i.e.

$$\rho_*(ij) \begin{bmatrix} I_0(1) \\ I_0(2) \\ I_0(3) \end{bmatrix} = R(ij) \begin{bmatrix} I_0(1) \\ I_0(2) \\ I_0(3) \end{bmatrix}$$

We have the following result.

$$R(12) = I + \begin{bmatrix} -\mu_1(1-\mu_2) & \mu_1(1-\mu_1) & 0 \\ (1-\mu_2) & -(1-\mu_1) & 0 \\ 0 & 0 & 0 \end{bmatrix}$$

$$R(23) = I + \begin{bmatrix} 0 & 0 & 0 \\ 0 & -\mu_2(1-\mu_3) & \mu_2(1-\mu_2) \\ 0 & (1-\mu_3) & -(1-\mu_2) \end{bmatrix}$$

$$R(13) = I + \begin{bmatrix} -\mu_1(1-\mu_3) & 0 & \mu_1(1-\mu_1) \\ (1-\mu_2)(1-\mu_3) & 0 & -(1-\mu_1)(1-\mu_2) \\ 1-\mu_3 & 0 & -(1-\mu_1) \end{bmatrix}$$

$$R(01) = I + \begin{bmatrix} \mu_0\mu_1 - 1 & 0 & 0 \\ \mu_0(\mu_2 - 1) & 0 & 0 \\ \mu_0(\mu_3 - 1) & 0 & 0 \end{bmatrix}$$

$$R(02) = I + \begin{bmatrix} 0 & \mu_1 - 1 & 0 \\ 0 & \mu_0\mu_2 - 1 & 0 \\ 0 & \mu_0(\mu_3 - 1) & 0 \end{bmatrix}$$

$$R(03) = \mu_4 \left[I + \begin{bmatrix} 0 & 0 & \mu_1 - 1 \\ 0 & 0 & \mu_2 - 1 \\ 0 & 0 & \mu_0\mu_3 - 1 \end{bmatrix} \right]$$

where I stands for the identity matrix and

$$\mu_j = e(\lambda_j) = \exp(2\pi i \lambda_j) \qquad 0 \leq j \leq 4$$

Any five of the six above matrices above generate the monodromy group of the equation $F_1(\lambda)$.

<u>Remark 1)</u>: If $\mu_i\mu_j \neq 1$ then $R(ij)$ is semi-simple with eigen values proportional to $1, 1, \mu_i\mu_j$. If $\mu_i\mu_j = 1$, then $R(ij)$ has Jordan normal form proportional to $\begin{bmatrix} 1 & 1 & 0 \\ 0 & 1 & 0 \\ 0 & 0 & 1 \end{bmatrix}$

<u>Invariant Hermitian matrix</u>: Let $\Gamma(\lambda)$ be the group generated by the $R(ij)$'s. Then there is, up to multiplicative constants, a unique $\Gamma(\lambda)$-invariant Hermitian matrix $H(\lambda)$ such that

$$^t\bar{X}H(\lambda)X = H(\lambda) \quad \text{for} \quad X \in \Gamma(\lambda)$$

Checking the action of the generators $R(ij)$, we have

$$H(\lambda) = \begin{bmatrix} 2\mathrm{Re}\,\dfrac{\sqrt{\mu_4}}{1-\bar{\mu}_1} & \mu_1\sqrt{\mu_4} & \mu_1\mu_2\sqrt{\mu_4} \\ \bar{\mu}_1\sqrt{\mu_4} & 2\mathrm{Re}\,\dfrac{\sqrt{\mu_4}}{1-\bar{\mu}_2} & \mu_2\sqrt{\mu_4} \\ \bar{\mu}_1\bar{\mu}_2\sqrt{\mu_4} & \bar{\mu}_2\sqrt{\mu_4} & 2\mathrm{Re}\,\dfrac{\sqrt{\mu_4}}{1-\bar{\mu}_3} \end{bmatrix}$$

One can check that if $0 < \lambda_j < 1$ ($0 \leq j \leq 4$) then the Hermitian matrix $H(\lambda)$ is of signature $(2,1)$ and that if $\lambda_0 = \lambda_1 = \lambda_2 = \lambda_3 = 3/4$ and $\lambda_4 = 0$ (cf. Remark (2) at the end of the previous paragraph) then $H(\lambda)$ degenerates. This is a special case of a theorem in [Ter 1].

<u>Remark</u> 2): The above calculation for the system F_1 works also for the equation F_D in N variables without any extra difficulty because it has the following integral representation

$$\int (t-x^0)^{\lambda_0-1}(t-x^1)^{\lambda_1-1}\ldots(t-x^{N+2})^{\lambda_{N+1}-1}\,dt$$

where $\lambda_0 + \ldots + \lambda_{N+2} = N+1$. Letting $N = 1$ we have the promised monodromy representation of the HGDE.

We have obtained generators of the monodromy group. But we do not know enough about the monodromy group itself. In the 27 monodromy groups of the systems in the PTMD-theorem (refer to them as <u>PTMD-groups</u>), there are arithmetic and non-arithmetic groups(cf. [D-M]). The following two cases have been studied arithmetically.

__Case__ (1) $\lambda_j = 3/5$ $(0 \leq j \leq 4)$: Put

$\mu = \exp(6\pi i/5)$
$O =$ the ring of integers in $\mathbf{Q}(\mu)$
$H = \begin{bmatrix} 1 & 0 & 0 \\ 0 & 1 & 0 \\ 0 & 0 & (1-\sqrt{5})/2 \end{bmatrix} \quad \varepsilon\ GL(3,O)$

$\Gamma = \{X \in GL(3,O) \mid {}^t\bar{X}HX = H\}$
$\Gamma((1-\mu)) = \{X \in \Gamma \mid X \equiv 1 \mod(1-\mu)\}$: the principal congruence subgroup of Γ with respect to the ideal $(1-\mu)$
Γ' : projectification of Γ
$\Gamma'((1-\mu))$: projectification of $\Gamma((1-\mu))$

__Fact__ ([Y-Y]): There is a matrix $Q \in GL(3,O)$ such that

${}^tQH(\lambda)Q = (\text{const.})H$
$Q^{-1}\Gamma(\lambda)Q = \Gamma((1-\mu))$
$\Gamma'/\Gamma'((1-\mu)) \cong S_5$ (symmetric group)

__Case__ (2) $\lambda_j = 2/3$ $(0 \leq j \leq 3)$, $\lambda_4 = 1/3$: Put

$\mu = \exp(2\pi i/3)$
$H = \begin{bmatrix} 1 & 0 & 0 \\ 0 & 1 & 0 \\ 0 & 0 & -1 \end{bmatrix}$

O, Γ, $\Gamma((1-\mu))$, Γ',... : defined as above.

__Fact__: There is a matrix $Q \in GL(3,O)$ such that

${}^tQH(\lambda)Q = (\text{const.})H$
$Q^{-1}\Gamma(\lambda)Q = \Gamma((1-\mu))$
$\Gamma'/\Gamma'((1-\mu)) \cong S_4$ (symmetric group)

__Note__: This case, first treated by E.Picard, has been deeply studied by R-P.Holzapfel ([Hol]) and H.Shiga ([Shg 2]).

Chapter 11 Reflection Groups

An automorphism g of a manifold M is called a <u>reflection</u> if g leaves a subvariety (called the <u>mirror</u> of g) of codimension one pointwise fixed and if the order of g is finite and not equal to 1. A <u>reflection group</u> is a group generated by reflections.

We have already encountered many reflections and reflection groups.

1) Let (X,S,b) be an orbifold and M be a uniformization. The element of Aut(M) corresponding to a loop around a non-singular part of S is a reflection.

2) A singular point of S is thus related to a (non-cyclic) reflection group in Aut(M).

3) The examples of orbifolds in § 10.2 covered by $\mathbb{C}^2$ are the quotients by reflection groups in E(2). They are called crystallographic reflection groups.

4) The PTMD-groups are reflection groups in $Aut(\mathbb{B}_2)$.

5) The covering groups of the orbifolds found by Höfer, Hirzebruch, and Hunt (§ 10.3) are also reflection groups in $Aut(\mathbb{B}_2)$.

6) The line arrangements in § 10.3 are derived from some reflection subgroups of the unitary groups.

In this chapter, we discuss several kinds of reflection groups. In § 11.1, fundamental definitions and classical facts for unitary reflection groups are stated. In § 11.2, 2-dimensional unitary reflection groups are presented. Many explanations of such groups known so far are not so easy to understand. We give here a new explanation by using illustrations of orbifolds which jump into eyes. In §11.3, 2-dimensional parabolic reflection groups, which are not classical, are studied. We list up all of them by using illustrations again. In § 11.4, we present 3-dimensional unitary reflection groups, which define some line arrangements in P_2. In each case, the groups we treat exhaust all such groups. But to show they are complete, one needs case by case checking; It is not difficult but tedious so it is omitted as in many books.

§ 11.1 Unitary Reflection Groups

Let $GL(n,\mathbb{C})$ be the general linear group acting on $\mathbb{C}^n$ ($n \geq 2$). A reflection in $GL(n,\mathbb{C})$ is characterized by the property that all but one of the eigenvalues are equal to 1 and the remaining one is a root of unity not equal to 1. A finite subgoup of $GL(n,\mathbb{C})$ generated by reflections is called a <u>unitary reflection group</u>, because any finite subgroup of $GL(n,\mathbb{C})$ is conjugate to a subgroup of the unitary group $U(n)$. We have the following exact sequence

$$1 \to GL(1,\mathbb{C}) \to GL(n,\mathbb{C}) \xrightarrow{\pi} PGL(n,\mathbb{C}) \to 1$$

For a subgroup G of $GL(n,\mathbb{C})$, the image $\pi(G)$ is called the <u>projectivized group</u> (or collineation group), and it acts on the projective space P_{n-1}. If a unitary reflection group is conjugate to a subgroup of $GL(n,\mathbb{R})$ then it is called a <u>(finite) Coxeter group</u>. A subgroup G of $GL(n,\mathbb{C})$ is said to be <u>imprimitive</u> if $\mathbb{C}^n$ is a nontrivial direct sum of vector spaces which are permuted by G, and is said to be <u>primitive</u> if it is not imprimitive.

A typical example of a primitive reflection group is the symmetric group S_{n+1}. It acts on $\mathbb{C}^{n+1}$ as permutations of $n+1$ coordinates $x_1, \ldots, x_{n+1}$. The permutation of two coordinates x_i and x_j corresponds to the reflection which fixes the hyperplane $\{x_j = x_i\} \subset \mathbb{C}^{n+1}$. Restricting this action to the invariant hyperplane $\mathbb{C}^n \cong \{x_1 + \ldots + x_{n+1} = 0\} \subset \mathbb{C}^{n+1}$ we get a faithful representation of S_{n+1} into $GL(n,\mathbb{R})$. This group is called the <u>Coxeter group of type</u> A_n. In this case, the projectivized group is isomorphic to A_n.

<u>Irreducible imprimitive groups</u> $G(m,p,n)$: For an integer $m \geq 2$, put $\theta = \exp 2\pi i/m$ and define a group $G(m,m,n) \subset U(n)$ to be the group of transformations of the form

$$(\sigma, a): \quad x_i' = \theta^{a_i} x_{\sigma(i)} \quad i = 1, \ldots, n$$

where $\sigma \in S_n$, $a_i \in \mathbb{Z}$ and $a = (a_1, \ldots, a_n)$ satisfies

$$\Sigma a_i \equiv 0 \quad \mod m.$$

Reflections in $G(m,m,n)$ are of the form

$$x_i' = \theta^a x_j$$
$$x_j' = \theta^{-a} x_i$$
$$x_k' = x_k \quad (k \neq i,j)$$

for i,j ($i \neq j$), and the group is generated by these reflections.

For an integer $m \geq 2$ and a divisor p ($1 \leq p \leq m$) of m, define a group $G(m,p,n) \subset U(n)$ to be the group generated by $G(m,m,n)$ and reflections of the form

$$x_i' = \theta^p x_i$$
$$x_k' = x_k \quad (k \neq i)$$

Any element of $G(m,p,n)$ is a transformation of the form (σ,a) which this time does not satisfy the condition $\Sigma \, a_i \equiv 0$ mod m. We have the following two split exact sequences

$$1 \to (\mathbb{Z}/m\mathbb{Z})^{n-1} \to G(m,m,n) \to S_n \to 1$$
$$a \mapsto (1,a)$$
$$(\sigma,a) \mapsto \sigma$$

$$1 \to G(m,m,n) \to G(m,p,n) \to \mathbb{Z}/q\mathbb{Z} \to 1$$
$$(\sigma,a) \mapsto \Sigma \, a_i/p \quad \text{mod} \quad q$$

where $pq = m$. In some cases the group $G(m,p,m)$ can be defined over the reals:

$G(2,1,n)$: Coxeter group of type B_n

$G(2,2,n)$: Coxeter group of type D_n

$G(m,m,2)$: Dihedral group of order $2m$

Other than these two kinds of examples, there are only finite numbers of irreducible unitary reflection groups, all of which are primitive. They are called <u>exceptional</u> unitary reflection groups. One can find a list of these in many places (see for example [S-T]). It is known that if n is greater than 8 then there are no exceptional reflection groups. Two and three dimensional exceptional groups are discussed in the following paragraphs.

Let R be the polynomial ring in n indeterminates over $\mathbb{C}$, and G be a finite subgroup of $GL(n,\mathbb{C})$. The group G acts on the ring R by

$$gP(x) = P(g^{-1}(x)) \qquad P(x) \in R, \qquad g \in G.$$

An element of R is called an <u>invariant</u> of G if it is invariant under the action of G. An element P of R is called an <u>anti-invariant</u> of G if

$$gP(x) = (\det g)^{-1} P(x) \qquad g \in G.$$

All the invariants form a ring which is called <u>the ring of invariants</u>; and is denoted by R^G. It is well known that the ring R^G is generated by finite number of homogeneous invariants (i.e. invariants which are homogeneous polynomials). There are many beautiful theorems about the invariants of reflection groups. The basic result is the following.

<u>Theorem</u> (Chevalley): A finite group $G \subset GL(n,\mathbb{C})$ is a reflection group if and only if the ring R^G is generated by n algebraically independent homogeneous invariants. (They are called a set of <u>fundamental invariants</u>.)

<u>Theorem</u>: Let $G \subset U(n)$ be a unitary reflection group and let D be the product of linear forms which define the mirrors of the reflections in G. Then the set of anti-invariants of G coincides with DR^G. If $I_1, \ldots, I_n$ denote fundamental invariants, then we have

$$D = c \det(\partial I_i / \partial x_j)$$

for some non-zero complex number c.

We can translate these into the language of geometry:

<u>Theorem</u>: A finite group $G \subset GL(n,\mathbb{C})$ is a reflection group if and only if the factor space $\mathbb{C}^n/G$ is isomorphic to $\mathbb{C}^n$ (in particular, it is non-singular at the origin).

Theorem: Let $G \subset U(n)$ be a unitary reflection group and let $p: \mathbb{C}^n \to \mathbb{C}^n/G$ be the natural projection. Then the set of critical points of p is the union of the mirrors of the reflections in G.

Remark: The set of degrees $\{d_1,\ldots,d_n\}$ of any set of fundamental invariants are uniquely determined by G. They satisfy

$$\prod d_i = \text{order } |G| \text{ of } G$$

$$\sum (d_i - 1) = \text{number of reflections in } G$$

If the reader is interested in the proof and more detailed results, see the above reference or [Br].

§ 11.2 Unitary Reflection Groups of Dimension 2

A finite subgroup of $PGL(2,\mathbb{C})$ is isomorphic to one of the following groups:

1) cyclic groups
2) symmetric group S_3
3) polyhedral group $<p,q,r>$ having the presentation

$$A^p = B^q = C^r = ABC = 1 \; ,$$

where the triad of integers p,q,r greater than 1 satisfies

$$1/p + 1/q + 1/r > 1 \; .$$

The order is $2s$ where s is given by

$$1/s = 1/p + 1/q + 1/r - 1.$$

All the polyhedral groups are listed below.

$$\begin{array}{lll} <s,2,2> : \text{dihedral group} & s = s \\ <3,3,2> : \text{tetrahedral group} & s = 6 \\ <4,3,2> : \text{octahedral group} & s = 12 \\ <5,3,2> : \text{icosahedral group} & s = 30 \end{array}$$

For a polyhedral group $<p,q,r>$, there is a <u>binary polyhedral group</u> $<p,q,r>_1$ having the presentation:

$$A^p = B^q = C^r = ABC.$$

Its order is 4s. There is also an <u>m-binary polyhedral group</u> $<p,q,r>_m$ of order 4ms having the presentation

$$A^p = B^q = C^r = ABC = Z^m, \quad [Z,A] = [Z,B] = [Z,C] = 1$$

(which implies $Z^{2m} = 1$) and admitting the exact sequence

$$\text{(ES)} \quad 1 \to \mathbf{Z}/2m\mathbf{Z} \to <p,q,r>_m \to <p,q,r> \to 1$$
$$\phantom{\text{(ES)} \quad 1 \to \mathbf{Z}/2m\mathbf{Z}} \quad j \mapsto Z^j$$

Let G' be a finite subgroup of $PGL(2,\mathbf{C})$. There are finitely many unitary reflection groups G whose projectivization $\pi(G)$ is G'.

1) If G' is cyclic then G is reducible, that is, G is conjugate to

$$\left\langle \begin{pmatrix} \exp 2\pi i/m & 0 \\ 0 & 1 \end{pmatrix}, \begin{pmatrix} 1 & 0 \\ 0 & \exp 2\pi i/n \end{pmatrix} \right\rangle \cong \mathbf{Z}/m\mathbf{Z} \times \mathbf{Z}/n\mathbf{Z}$$

for some integers m and n. Here $<a,b,...>$ stands for the group generated by $a, b, \ldots$.

2) If G' is S_3 then G is the Coxeter group A_2 as we studied in § 11.1.

3) If G' is a polyhedral group $<p,q,r>$ then there is a __maximal__ unitary reflection group, which is isomorphic to $<p,q,r>_s$. Every reflection group whose projectivization is G' is a normal subgroup of this maximal group. It is generated by three reflections R_1, R_2 and R_3 with the following relations:

$$[R_1R_2R_3, R_j] = 1 \qquad (j = 1,2,3)$$
$$R_1^p = R_2^q = R_3^r = 1$$

To show this presentation, we change the above generator A, B, C and Z of $<p,q,r>_s$ into the generators R_1, R_2 and R_3 by putting

$$R_1 = Z^{s/p}A, \quad R_2 = Z^{s/q}B, \quad R_3 = Z^{s/r}C.$$

(The center is generated by $Z = R_1R_2R_3$.)

If G is the maximal reflection group $<p,q,r>_s$ then the image of the mirrors in G under the projection $\mathbb{C}^2 \to \mathbb{C}^2/G \cong \mathbb{C}^2$ is the union of three lines passing through the same point. This can be seen from the presentation above (using R_1, R_2 and R_3 as generators) and the following remark, which also explains a geometric meaning of the exact sequence (ES) for m = s.

__Remark__: Let $L_1,\ldots,L_r$ be r lines in $\mathbb{C}^2$ passing through the origin. Then the fundamental group $\Pi_1(\mathbb{C}^2 - \cup L_j)$ has the following presentation

$$[R_1\ldots R_r, R_j] = 1 \qquad (j = 1,\ldots,r)$$

Proof: Let L_0 be another line passing through the origin O; let H be a line not passing through O but intersecting with lines $L_0, L_1, \ldots, L_r$ at points $a, p_1, \ldots, p_r$, respectively; and let R_j ($1 \leq j \leq r$) be loops in $H' = H - \bigcup_{j=1}^{r} \{p_j\}$ starting and ending at a defined as follows:

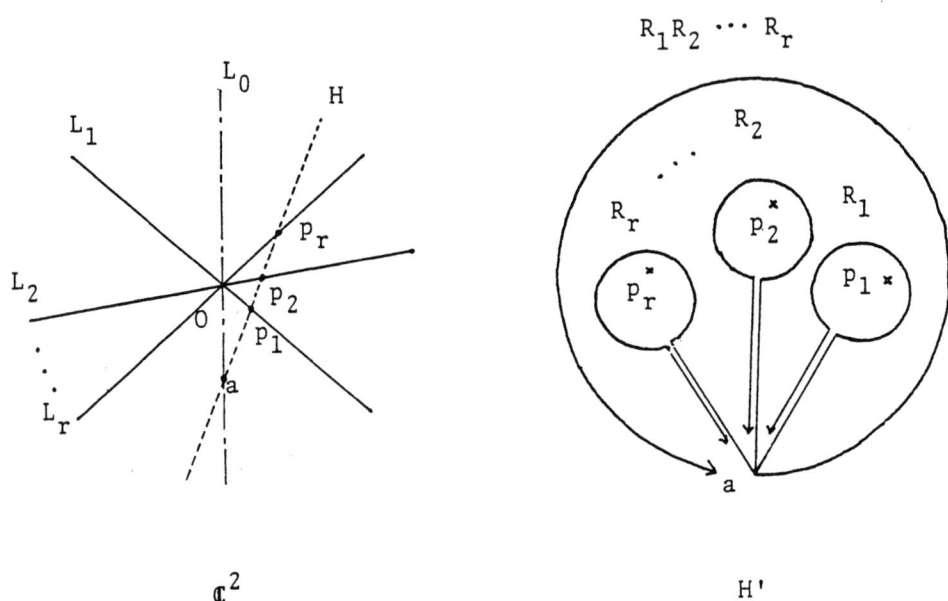

These R_j's generate the fundamental group $\Pi(\mathbb{C}^2-L, a)$, where $L = \bigcup_{j=1}^{r} L_j$. Let finally R_0 be the loop in $L_0 - \{0\}$, going around the origin once in the positive direction, defined by

$$[0,1] \ni t \to e^{2\pi i t} a \in L_0 - \{0\},$$

which generates $\Pi_1(L_0 - \{0\}, a) \cong \mathbb{Z}$. Notice that R_0 equals $R_1 \ldots R_r$ in $\Pi_1(\mathbb{C}^2 - L, a)$. (See the above picture.) Notice also that R_0 belongs to the centre of $\Pi_1(\mathbb{C}^2 - L, a)$, in particular, we have

$$[R_0, R_j] = 1 \qquad (j = 1, \ldots, r).$$

Indeed for any loop

$$\gamma : [0,1] \to \mathbb{C}^2 - L \qquad \gamma(0) = \gamma(1) = a$$

in $\mathbb{C}^2-L$, there is a family of loops $F_s: [0,2] \to \mathbb{C}^2-L$ with a parameter s ($0 \leq s \leq 1$) given by

$$F_s(t) = \begin{cases} \gamma(t) & 0 \leq t \leq s \\ e^{2\pi i(t-s)}\gamma(s) & s \leq t \leq s+1 \\ \gamma(t-1) & s+1 \leq t \leq 2 \end{cases}$$

which connects $F_0 = R_0\gamma$ and $F_1 = \gamma R_0$.

On the other hand, consider the natural projection $q: \mathbb{C}^2-\{0\} \to \mathbb{P}_1$, and denote by $b, x_1, \ldots, x_r$ the images of lines L_0, $L_1, \ldots, L_r$, respectively. Then we have the fibration (a restriction of the above projection):

$$q: \mathbb{C}^2 - L \to \mathbb{P}_1 - x \qquad (x = \cup_{i=1}^r \{x_i\})$$

with the typical fibre L_0; which induces the exact sequence

$$1 \to \Pi_1(L_0-\{0\},a) \to \Pi_1(\mathbb{C}^2-L,a) \xrightarrow{q_*} \Pi_1(\mathbb{P}_1-x,b) \to 1.$$

(Compare this exact sequence with (ES).) Since $\Pi_1(\mathbb{C}^2-L,a)$ is generated also by $R_0, R_1, \ldots, R_{r-1}$, any word W in $R_1, \ldots, R_r$ can be written as $W = R_0^s W_1$ in $\Pi_1(\mathbb{C}^2-L,a)$ where $s \in \mathbb{Z}$ and W_1 is a word in $R_1, \ldots, R_{r-1}$. If the word W represents 1 in $\Pi_1(\mathbb{C}^2-L,a)$, then we have $q_*W_1 = 1$ in $\Pi_1(\mathbb{P}_1-x,b)$, which is the free group generated freely by $q_*R_1, \ldots, q_*R_{r-1}$; This implies that the word W_1 itself is trivial. Thus we have proved that $\Pi_1(\mathbb{C}^2-L,a)$ is generated by $R_1,\ldots,R_r$; and that $[R_1\ldots R_r, R_j] = 1$ ($j = 1,\ldots,r$) generate the relations.

For the dihedral group $<s,2,2>$ the maximal unitary reflection group is given by

$$<s,2,2>_s = G(2s,2,2).$$

Indeed, since

$$G(m,p,2) = \left\langle \begin{pmatrix} 0 & 1 \\ 1 & 0 \end{pmatrix}, \begin{pmatrix} 0 & \theta \\ \theta^{-1} & 0 \end{pmatrix}, \begin{pmatrix} \theta^p & 0 \\ 0 & 1 \end{pmatrix} \right\rangle$$

where $\theta = \exp 2\pi i/m$, we have

$$\pi G(m,p,2) \cong <\theta^{(2,p)}> \rtimes \mathbb{Z}/2\mathbb{Z}$$

where $(2,p)$ stands for the greatest common divisor of 2 and p. Fundamental invariants of $G(m,p,2)$ are given by

$$x^m + y^m \text{ and } (xy)^{m/p}.$$

Now we are going to present all the 2-dimensional unitary reflection groups G. Instead of listing various data (e.g. orders, fundamental invariants, etc) or generators and relations of G, we describe G by illustrating the curve $S = \cup S_j$ and the weights b_j of the corresponding orbifolds $\underline{X} = (X,S,b)$ where $X = \mathbb{C}^2/G$ ($\cong \mathbb{C}^2$) and S is the image of the union of the mirrors of the reflections in G.

Conventions:

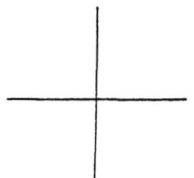
: a non-singular curve with weight b

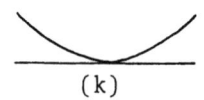
: two non-singular curves meeting transversely at a point

(k)
: two non-singular curves which are tangent with multiplicity k

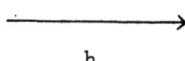
(a,b)
: a cusp of type (a,b) i.e. an irreducible curve locally equivalent to $x^a + x^b = 0$

$\xrightarrow{}$
h
: cyclic branched cover of degree h

((k))
: the <u>identification number</u> of the group in the Shephard-Todd classification ([S-T])

Before listing pictures of orbifolds, we explain how to read them by giving the following examples of pictures of orbifolds: $X_j = \mathbb{C}^2/G_j$ ($j = 1,2,3$), where G_j's are reflection groups whose projectivized groups are equal to $<4,3,2>$.

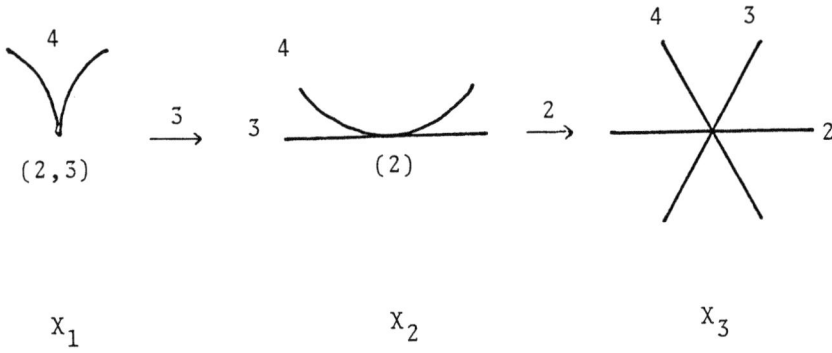

X_1 $\qquad\qquad\qquad X_2 \qquad\qquad\qquad X_3$

Orbifold X_1 is the triple covering of X_2 branching along the nonsingular curve in X_2 labelled 3; and orbifold X_2 is the double covering of X_3 branching along the curve in X_3 labelled 2. For each j ($j = 1,2,3$), we can choose systems of coordinates (x_j, y_j) on $X_j \cong \mathbb{C}^2$ so that the singular loci S_j of orbifolds X_j are given by

$$S_1 = \{ x_1^2 = y_1^3 \}, \quad S_2 = \{ y_2 = 0 \} \cup \{ x_2^2 = y_2 \}$$
$$S_3 = \{ x_3 = 0 \} \cup \{ y_3 = 0 \} \cup \{ x_3 = y_3 \};$$

that the projection $X_1 \to X_2$ is given by $(x_2, y_2) = (x_1, y_1^3)$ and that $X_2 \to X_3$ is given by $(x_3, y_3) = (x_2^2, y_2)$.

Orbifolds $\underline{X} = \mathbb{C}^2/G$

$\mathbb{Z}/m\mathbb{Z} \times \mathbb{Z}/n\mathbb{Z}$ A_2

$\overline{G} = \langle s, 2, 2 \rangle$, $s = s_1 s_2$

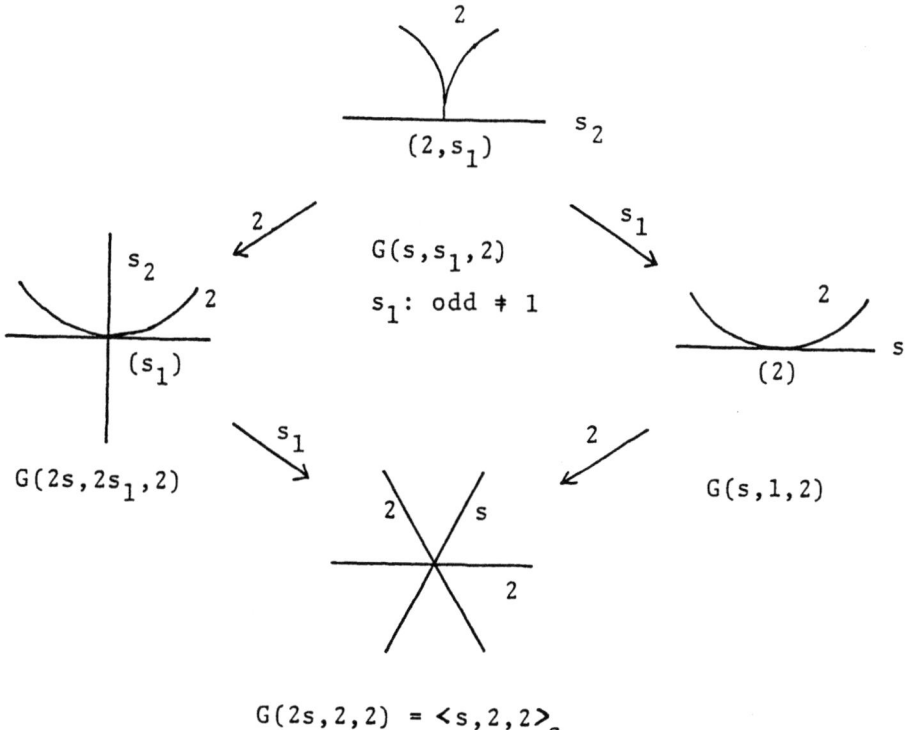

$G(2s, 2, 2) = \langle s, 2, 2 \rangle_s$

If $s_2 = 1$, we regard the curve labelled s_2 as being absent.

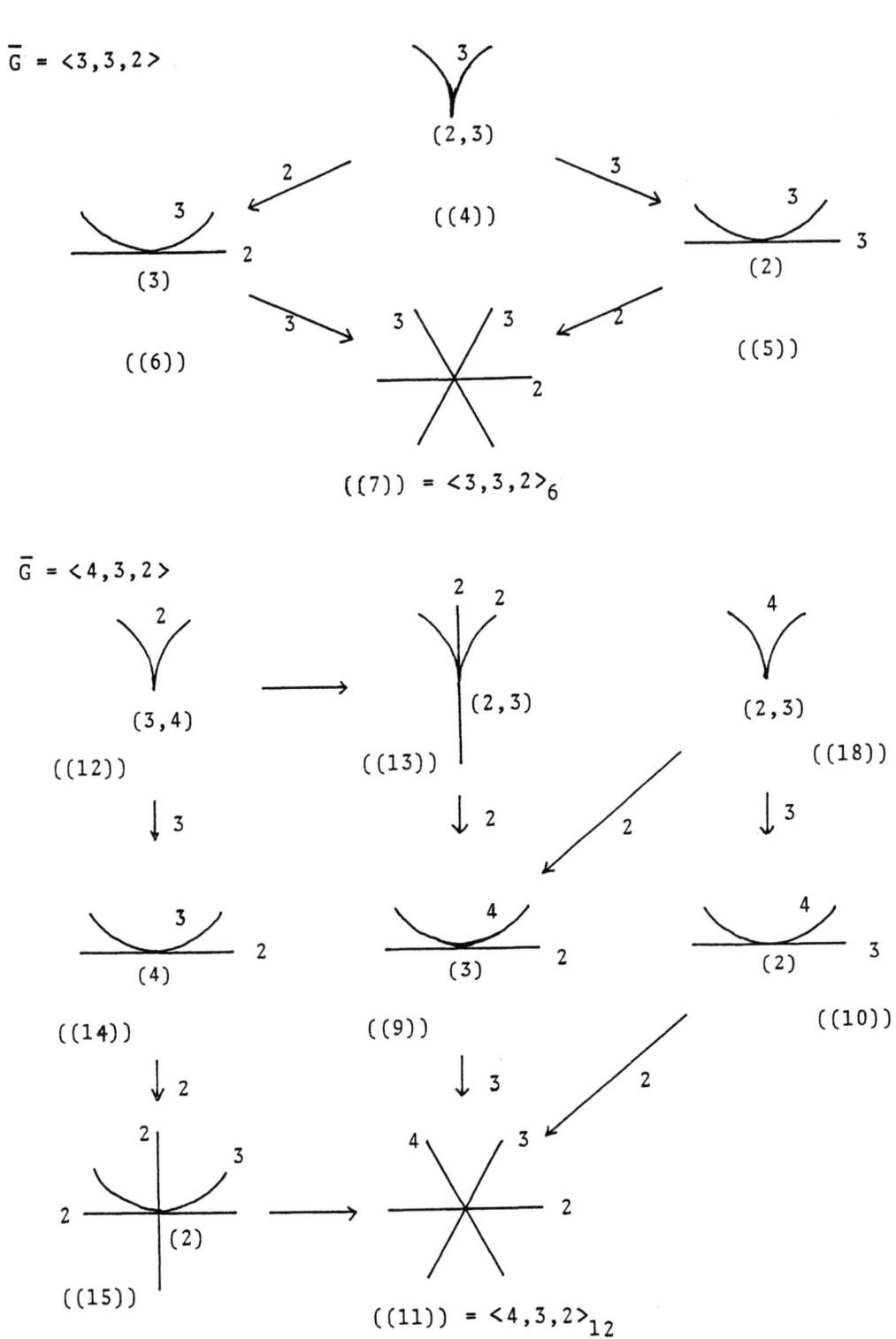

169

$\bar{G} = \langle 5,3,2 \rangle$

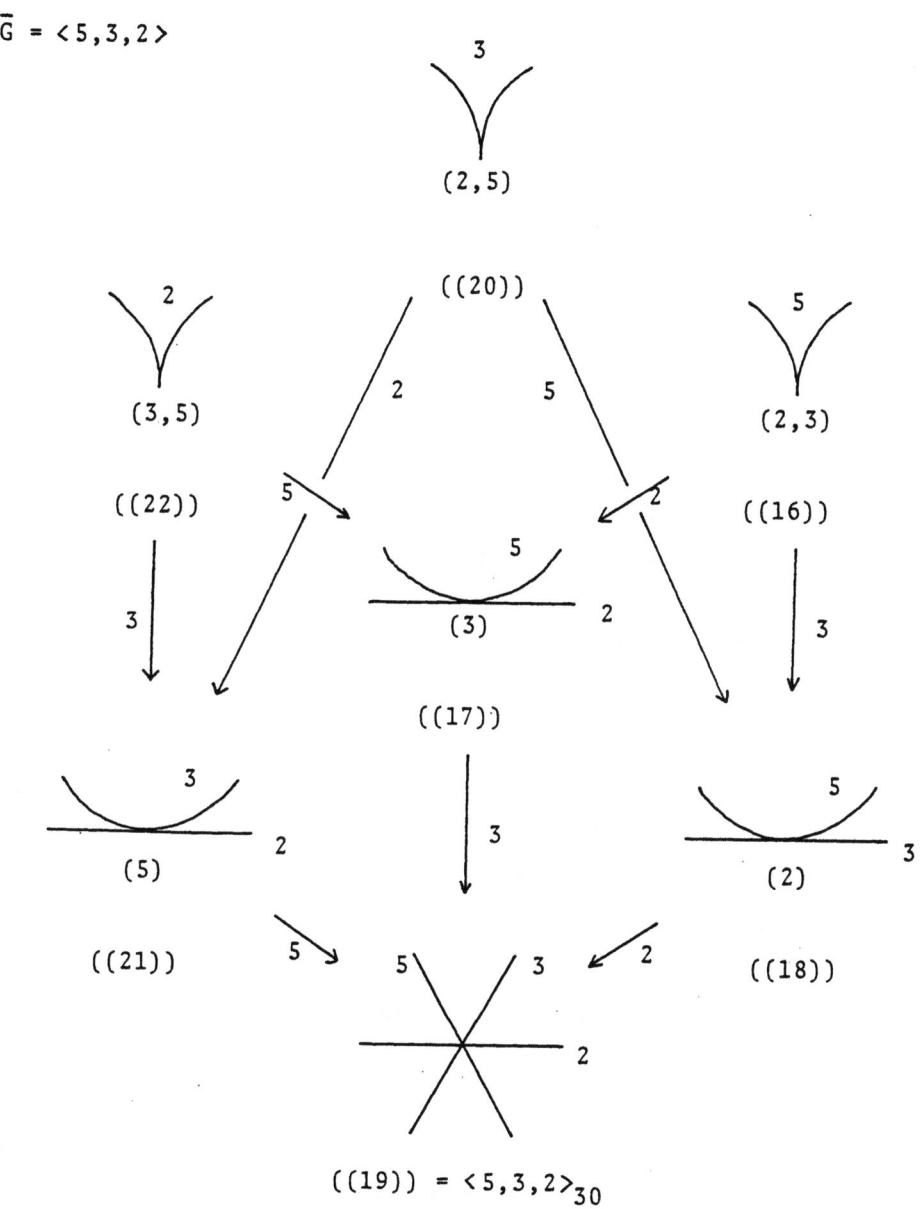

Let (X,S,b) be a 2-dimensional orbifold and let p be a point of S. If there is a neighbourhood U of p such that the orbifold $\underline{U} = (U, S \cap U, b|_S)$ admits a finite uniformization then, locally at p, the orbifold $\underline{U}$ must be equivalent to one of the orbifolds listed above.

§ 11.3 Parabolic Reflection Groups of Dimension 2

Let (X,S,b) be a 2-dimensional orbifold uniformized by the ball B_2 and let $\Gamma \subset \text{Aut}(B_2)$ be the covering transformation group. If Γ is not cocompact (i.e. if the quotient space is not compact), then corresponding to the cusps of Γ, the weight of some S_j is infinite. We shall study the cusps of Γ and shall find that the structure of a neighbourhood of such S_j is again very restricted. (The reader should notice that the following argument has some similarity to that of § 11.2)

We consider a boundary point of the ball

$$B_2 = \{ (1, z_1, z_2) \in \mathbb{P}_2 \mid |z_1|^2 + |z_2|^2 - 1 < 0 \}$$

in $\mathbb{P}_2$. As in the single variable case, it is convenient to transform B_2 (projectively) so that the boundary point is at infinity. We let

$$D = \{ (z, u, 1) \in \mathbb{P}_2 \mid \text{Im } z - |u|^2 > 0 \} \quad (\cong \hat{B}_2)$$

and $\quad p = (1,0,0) \in \partial D \subset \mathbb{P}_2$.

We consider the subgroup P of automorphisms of D fixing the boundary point p of the following form:

$$P = \{ ((\mu, a, r)) \mid \mu \in U(1), a \in \mathbb{C}^1, r \in \mathbb{R} \}$$

where $((\mu,a,r))$ is the automorphism

$$\begin{bmatrix} 1 & 2\pi i\mu a & r+i|a|^2 \\ 0 & \mu & a \\ 0 & 0 & 1 \end{bmatrix}$$

This group is called the <u>parabolic subgroup of</u> Aut(D) corresponding to the boundary point p. Let us define a homomorphism π of P onto the 1-dimensional Euclidean motion group $E(1) = U(1) \ltimes \mathbb{C}^1$, acting on $\mathbb{C}^1$, by forgetting the first line and the first column:

$$\pi : P \to E(1)$$
$$((\mu,a,r)) \mapsto \begin{pmatrix} \mu & a \\ 0 & 1 \end{pmatrix}$$

and define two normal subgroups of P:

$$P_1 = \{ ((1,a,r)) \in P \}$$
$$Z = \{ ((1,0,r)) \in P \} : \text{the centre of } P$$

Then we have the following commutative diagram of exact sequences:

$$\begin{array}{ccccccccc}
 & & 1 & & 1 & & 1 & & \\
 & & \downarrow & & \downarrow & & \downarrow & & \\
1 & \to & Z & \to & P_1 & \to & \mathbb{C}^1 & \to & 1 \\
 & & \downarrow & & \downarrow \pi & & \downarrow & & \\
1 & \to & Z & \to & P & \to & E(1) & \to & 1 \\
 & & \downarrow & & \downarrow & & \downarrow & & \\
 & & 1 & \to & U(1) & \to & U(1) & \to & 1 \\
 & & & & \downarrow & & \downarrow & & \\
 & & & & 1 & & 1 & &
\end{array}$$

where the first row is given by the homomorphisms

$r \mapsto ((1,0,r))$ and $((1,a,r)) \mapsto a$; and the third column is given by $a \mapsto \begin{pmatrix} 1 & a \\ 0 & 1 \end{pmatrix}$ and $\begin{pmatrix} \mu & a \\ 0 & 1 \end{pmatrix} \mapsto \mu$.

<u>Definition</u>: A discrete subgroup Γ of P is said to be of <u>locally finite covolume</u> if for a positive number N, the factor space of

$$\{ (z,u) \in \mathbb{C}^2 \mid \text{Im } z - |u|^2 > N \} \subset D$$

by Γ has a finite volume with respect to the $\text{Aut}(D)$-invariant metric

$$\frac{(\text{Im } z - |u|^2)|du|^2 + |dz/2i - u\, du|^2}{(\text{Im } z - |u|^2)^2}$$

(The Poincaré-Bergmann metric).

Let Γ be a discrete subgroup of locally finite covolume. Then the center $Z(\Gamma)$ of Γ is isomorphic to $\mathbb{Z}$ and the image $\pi(\Gamma)$ is a 1-dimensional crystallographic group i.e. a cocompact discrete subgroup of $E(1)$. By Bieberbach's theorem, every crystallographic group is a finite extention of a <u>lattice</u> in $\mathbb{C}^1$ by a finite group ($\subset U(1)$) called <u>the point group</u>. Let $L(\Gamma)$ and $G(\Gamma)$ be the lattice and the point group of the crystallographic group $\pi(\Gamma)$ and put $\Gamma_1 = \Gamma \cap P_1$. Then we have the following commutative diagram of exact sequences:

$$\begin{array}{ccccccccc}
 & & 1 & & 1 & & 1 & & \\
 & & \downarrow & & \downarrow & & \downarrow & & \\
1 & \to & Z(\Gamma) & \to & \Gamma_1 & \to & L(\Gamma) & \to & 1 \\
 & & \downarrow & & \downarrow \pi & \downarrow & & \\
1 & \to & Z(\Gamma) & \to & \Gamma & \to & \pi(\Gamma) & \to & 1 \\
 & & \downarrow & & \downarrow & & \downarrow & & \\
 & & 1 & \to & G(\Gamma) & \to & G(\Gamma) & \to & 1 \\
 & & & & \downarrow & & \downarrow & & \\
 & & & & 1 & & 1 & &
\end{array}$$

If $G(\Gamma)$ is trivial i.e. $\Gamma = \Gamma_1$, $\pi(\Gamma) = L(\Gamma)$, then a smooth completion (at p) of the factor space D/Γ is obtained by adding the elliptic curve $\mathbb{C}/L(\Gamma)$. The self-intersection number of the curve in the space $(D/\Gamma) \cup (\mathbb{C}/L(\Gamma))$ is known to be

$$-4 \ \mathrm{vol}(\mathbb{C}/L(\Gamma))/ \ \mathrm{Min}\{ \ |r| \ ; \ ((1,0,r)) \in Z(\Gamma) \ \}.$$

<u>Note</u>: The argument of this paragraph so far can be generalized word for word to the n-dimensional case with the following changes:

$u \in \mathbb{C}^1 \quad \to \quad (u_1,\ldots,u_{n-1}) \in \mathbb{C}^{n-1}$

$(z,u) \quad \to \quad (z,u_1,\ldots,u_{n-1}) \in P_n$

$U(1) \to U(n-1), \quad\quad E(1) \to E(n-1)$

elliptic curve $\mathbb{C}/L(\Gamma) \to$ abelian variety $\mathbb{C}^{n-1}/L(\Gamma)$

self-intersection number

$\quad\quad\quad \to$ Riemann form $L(\Gamma) \times L(\Gamma) \to Z$ defined by

$(x,y) \to \mathrm{Im} \ {}^t x.y \ / \ \mathrm{Min}\{|r|;((1_{n-1},0,r)) \in Z(\Gamma)\}$

Suppose G is non-trivial. Since we have the exact sequence

$$1 \to L(\Gamma) \to \pi(\Gamma) \to G(\Gamma) \to 1$$

the lattice $L = L(\Gamma)$ is invariant under $G = G(\Gamma)$. Up to conjugacy there are four possible pairs (G,L) of G-invarianrt lattices L:

$G = \langle -1 \rangle \quad\quad L = Z + \tau Z \ (\tau \in H) \quad$ general lattice
$G = \langle \sqrt{-1} \rangle \quad\quad L = Z + \sqrt{-1} Z \quad\quad\quad$ square lattice
$G = \langle \omega \rangle \quad\quad L = Z + \omega Z \quad\quad\quad\quad$ hexagonal lattice
$G = \langle \sqrt{\omega} \rangle \quad\quad L = Z + \omega Z \quad\quad\quad\quad$ hexagonal lattice

where ω is a cube root of unity. In each case, the crystallographic group is the semi-direct product $L \rtimes G$ of the lattice L and the point group G ("semi-direct product" is defined in § 5.2). We shall call them

$$< 2,2,2,2;\tau >, < 4,4,2 >, < 3,3,3 >, < 6,3,2 >,$$

respectively. The group $< 2,2,2,2;\tau >$ has the presentation

$$A^2 = B^2 = C^2 = D^2 = ABCD = 1$$

and has the parameter τ. For each triad of integers p,q,r satisfing

$$1/p + 1/q + 1/r = 1$$

the group $< p,q,r >$ has the presentation

$$A^p = B^q = C^r = ABC = 1.$$

In these four cases the quotient space $\mathbb{C}/\pi(\Gamma)$ is rational (i.e. is a compact Riemann surface of genus 0).

A smooth completion (at p) of the factor space D/Γ is obtained by adding a finite number of rational curves (see [Y-H]), but in some cases (see below) it is done by adding only a point p.

An element $g \in \text{Aut}(D)$ is a reflection if it is of finite order and leaves the intersection of $D = \{(z,u) \in \mathbb{C}^2 | \text{Im } z - |u|^2 > 0\} \subset \mathbf{P}_2$ and a (complex) projective line passing through D pointwise fixed. A reflection $g \in P$ fixes a line passing through the point p. We have an analogue of Chevalley's theorem.

<u>Theorem</u>: Let Γ be a discrete subgroup of P of locally finite covolume. The space $D/\Gamma \cup \{p\}$ added by the point p (the Satake local completion of D/Γ) is non-singular if and only if Γ is a reflection group.

Proof: The following three facts imply the theorem.

i) (Mumford's criterion) If X is a two dimensional variety and X-{p} is smooth then X is non-singular at p if and only if there is a neighbourhood U of p in X such that the fundamental group $\Pi_1(U-\{p\})$ is trivial.

ii) ([Arm]) Let Γ be a properly discontinuous transformation group of a manifold M, and let X be the quotient space of M by Γ. Then the natural homomorphism $\Pi_1(M) \to \Pi_1(X)$ is surjective if and only if Γ is generated by elements each of which has a fixed point.

iii) (obvious) The set of fixed points in D of a non-trivial element of Aut(D) is either empty, a point or a line through D.

Remark: This proof also works for 2-dimensional Chevalley's theorem (§ 11.1).

Up to conjugacy there are finite number of reflection subgroups of P (two of them involve a parameter $\tau \in H$), which are called <u>parabolic reflection groups</u>. If Γ' is a crystallographic group with non-trivial point group, i.e.

$$< 2,2,2,2;\tau > \quad \text{or} \quad < p,q,r > \quad (1/p + 1/q + 1/r = 1),$$

then there is a <u>maximal</u> parabolic reflection group Γ such that $\pi(\Gamma) = \Gamma'$. Every parabolic reflection group is a normal subgroup of a maximal one. The corresponding maximal groups are denoted by

$$< 2,2,2,2;\tau >_\infty \quad \text{and} \quad < p,q,r >_\infty \quad (1/p + 1/q + 1/r = 1).$$

Each maximal group is generated by four or three reflections with the following relations. (Recall § 11.2 Remark.)

$< 2,2,2,2;\tau >_\infty$:

$$[R_1 R_2 R_3 R_4, R_j] = 1, \quad R_j^2 = 1 \quad (j = 1,2,3,4)$$

$< p,q,r >_\infty$ $(1/p + 1/q + 1/r = 1)$:

$$[R_1R_2R_3, R_j] = 1 \quad (j = 1,2,3) \qquad R_1^p = R_2^q = R_3^r = 1$$

Instead of listing every parabolic reflection group Γ, we illustrate the corresponding orbifold $\underline{X} = (X,S,b)$ where $X = D/\Gamma$ and S is the image of the union of the mirrors of the reflections in Γ. The space X is a contractible domain (in $\mathbb{C}^2$) minus a point (the image of the point p), which is denoted by a small circle "o". We use the same conventions as in the previous paragraph.

Orbifolds $\underline{X} = D/\Gamma$

$\pi(\Gamma) = <2,2,2,2;\tau>$

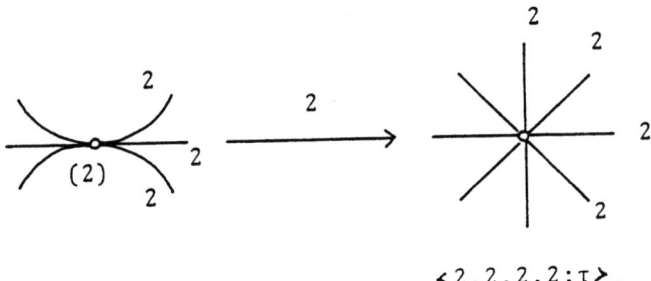

$<2,2,2,2;\tau>_\infty$

$\pi(\Gamma) = \langle 3,3,3 \rangle$

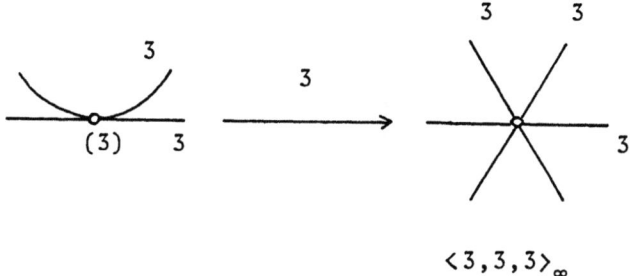

$\pi(T) = \langle 4,4,2 \rangle$

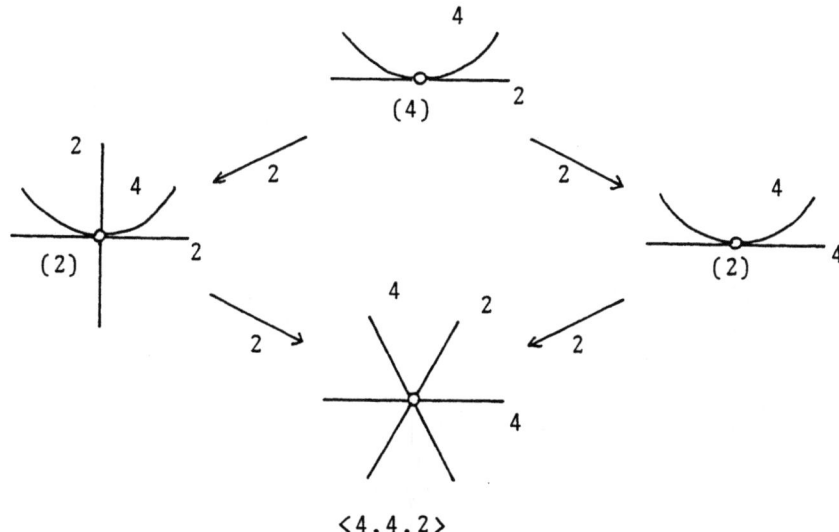

$\pi(\Gamma) = \langle 6,3,2 \rangle$

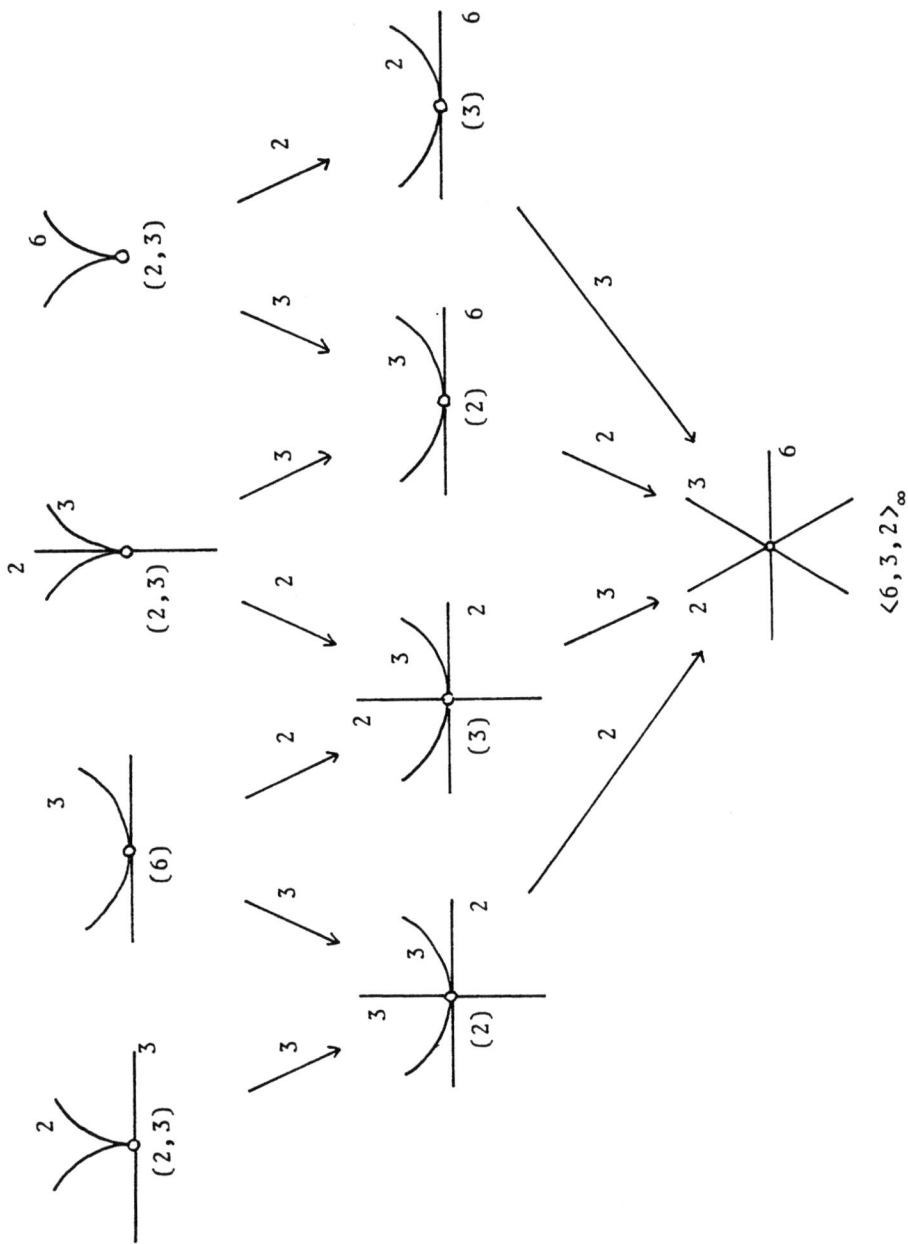

Remarks:1) For every 2-dimentional parabolic reflection group, fundamental invariants are known([Y-H]).

2) The parabolic reflection groups in higher dimensions are studied in [Yos 1] and [Kg].

3) Let $\Gamma \subset \text{Aut}(B_2)$ be a PTMD-group, and let $p \in B_2$ (resp. ∂B_2) be a fixed point of Γ, then the isotropy subgroup of Γ at p (the subgroup of Γ which fixes p) is a maximal unitary (resp. parabolic) reflection group.

§ 11.4 Line Arrangements Defined by Unitary Reflection Groups of Dimensions 3 and 4

We encountered in Chapter 10 several line arrangements (Examples 1,...,9) of P_2. They are related to unitary reflection groups of dimension 3 (Examples 1,...,8) and dimension 4 (Example 9).

Let G be a finite reflection group in $GL(3,\mathbb{C})$ acting on $\mathbb{C}^3$, and $A_1,\ldots,A_k$ be the mirrors of the reflections in G. Passing to the 2-dimentional projective space P_2, the mirrors define a line arrangement consisting of k lines $L_1,\ldots,L_k$.

This is called <u>the line arrangement defined by the unitary reflection group</u> G. If r_j denotes the number of reflections of order j in G, then we have

$$k = r_2 + r_3/2 + r_4/3 + \ldots$$

The complete quadrilateral (Example 1) is defined by the Coxeter group A_3.

The arrangements $A_3^0(m)$ ($m \geq 3$) and $A_3^3(m)$ ($m \geq 2$) (Examples 2 and 3, respectively) are defined by the imprimitive groups $G(m,m,3)$ and $G(m,p,3)$ ($p \neq m$), respectively.

The arrangements of Examples 4,...,8 are defined by the exceptional reflection groups. The following is a table of 3-dimensional exceptional unitary reflection groups, where the first column is the name of the group, the second is the registration number according to the Shephard-Todd classification, $|G|$ is the order of G, $|G'|$ is the order of the projectified group G', d_1, d_2, d_3 are the degrees of the fundamental invariants, and r_j and k are as above.

Name	No	$|G|$	$|G'|$	d_1, d_2, d_3	r_2	r_3	k
Icosahedral	23	120	60	2,6,10	15	0	15
Klein	24	336	168	4,6,14	21	0	21
Hesse	25	648	216	6,9,12	0	24	12
ext.-Hesse	26	1296	216	6,12,18	9	24	21
Valentiner	27	2160	360	6,12,30	45	0	45

Since the icosahedral arrangement is defined over the real numbers we can see it. In § 10.3 we include a picture of the icosahedron in $\mathbb{R}^3$ in which we can see the 15 planes A_j of symmetry. The F_4-arrangement (Example 9) is defined by the 4-dimensional Coxeter group F_4 as follows. As in 3-dimensional case, the 24 mirrors of the group F_4 define a hyperplane arrangement in $\mathbb{C}^4$. Passing to $\mathbb{P}_3$, this defines a plane arangement in $\mathbb{P}_3$. The F_4-arrangement in $\mathbb{P}_2$ is now defined as the restriction of the plane arrangement to any plane in the arrangement (any choice of plane produces the same line arrangement).

Chapter 12   Toward Finding New Differential Equations

In Chapter 10, we showed that for each line arrangement $A \subset P_2$ given in § 10.3, there is a Fuchsian differential equation defined on P_2 of rank 3 with regular singularities along A (the uniformizing differential equation of the orbifold). For complete quadrilateral arrangement, the equation is the Appell's system F_1, which was studied in § 10.5. In this chapter, we find differential equations for the arrangements given in Examples (2),...,(8). So this is a continuation of § 10.5. On the other hand, as we mentioned in § 9.1, it can be considered as a method for solving the Riemann problem in a very restricted case.

§ 12.1   Tensor Forms Associated to the Differential Equations

Let $x = (x^1, x^2)$ be a system of inhomogeneous coordinates on P_2 and consider an integrable system

$$E : \quad \frac{\partial^2 u}{\partial x^i \partial x^j} = p_{ij}^k(x) \frac{\partial u}{\partial x^k} + p_{ij}^0(x) u \qquad i,j = 1,2$$

in canonical form (i.e. $p_{im}^m(x) = 0$, $i = 1,2$). If $y = (y^1, y^2)$ is another set of inhomogeneous coordinates on P_2, then the coefficient of $\frac{\partial u}{\partial y^k}$ of the system E in canonical form with independent variables y^1, y^2 is given by

$$p^a_{bc}(x(y)) \frac{\partial x^b}{\partial y^i} \frac{\partial x^c}{\partial y^j} \frac{\partial y^k}{\partial x^a}$$

(Recall the Einstein convention). On the other hand, we already know (§ 7.4) and see soon later again that the p^0_{ij} can be expressed in terms of the p^k_{ij}'s ($i,j,k = 1,2$). Thus the tensor form

$$\omega'(E) = p^k_{ij}(x) dx^i \otimes dx^j \otimes \frac{\partial}{\partial x^k}.$$

determines the equation E.

Firstly we modify the tensor form $\omega'(E)$ to get a more symmetric expression. Let T, T^* and K be the tangent, the cotangent and the canonical bundle of P_2, respectively. The form $\omega'(E)$ is a (rational) section of the bundle $T^* \otimes T^* \otimes T$. Since we are working in <u>two-dimensional</u> space, we have the following isomorphism $T^* \cong T \otimes K$ by

$$p_1 dx^1 + p_2 dx^2 \longleftrightarrow (-p_2 \frac{\partial}{\partial x^1} + p_1 \frac{\partial}{\partial x^2}) \otimes (dx^1 \wedge dx^2).$$

This implies the isomorphism

$$T^* \otimes T^* \otimes T \cong T^* \otimes T^* \otimes T^* \otimes K^{-1}$$

If we put

$$p_{ij1}(x) = - p^2_{ij}(x), \quad p_{ij2}(x) = p^1_{ij}(x) \quad i,j = 1,2$$

then the section $\omega(E)$ of $T^* \otimes T^* \otimes T^* \otimes K^{-1}$ corresponding to the form $\omega'(E)$ is given by

$$\omega(E) = p_{ijk}(x) dx^i \otimes dx^j \otimes dx^k \otimes (dx^1 \wedge dx^2)^{-1}$$

Notice that the $p_{ijk}(x)$'s are __symmetric__ (i.e. $p_{ijk} = p_{jik} = p_{ikj}$) because of the conditions

$$p_{ij}^k(x) = p_{ji}^k(x) \qquad i,j,k = 1,2$$

and
$$p_{im}^m(x) = 0 \qquad i = 1,2.$$

Let V be a 3-dimensional vector space with coordinates $y = (y^0, y^1, y^2)$ and let $\pi: V - \{0\} \to P_2$ be the natural map given by

$$\pi : (y^0, y^1, y^2) \mapsto (x^1, x^2) = (y^1/y^0, y^2/y^0).$$

We now lift the form $\omega(E)$ by π. Put

$$\Omega(E) = \pi^* \omega(E)$$

and write it as

$$\Omega(E) = Q_{abc}(y) dy^a \otimes dy^b \otimes dy^c \otimes \gamma(y)^{-1}$$

where a, b and c run through $0, 1, 2$ and

$$\gamma(Y) = \pi^*(dx^1 \wedge dx^2)$$
$$= y^0 dy^1 \wedge dy^2 + y^2 dy^0 \wedge dy^1 + y^1 dy^2 \wedge dy^0.$$

The form $\Omega(E)$ has the following properties:
(0) $Q_{ijk}(y^0, y^1, y^2) = p_{ijk}(y^1/y^0, y^2/y^0)$ $\qquad i,j,k = 1,2$
(1) $Q_{abc}(y)$ is symmetric i.e. $Q_{abc} = Q_{acb} = Q_{bac}$

(2) $\Omega(E)$ is homogeneous of degree 0, i.e. the $Q_{abc}(y)$'s are homogeneous rational functions of degree 0

(3) $y^c Q_{abc}(y) = 0$ $a,b = 0,1,2$.

The last two properties comes from the fact that the form $\Omega(E)$ is the lift of a form on P_2.

Note: In general, a form Ω on V is the lift of a form on P_2 if and only if the Lie derivative and the interior product of Ω with respect to the vector field

$$y^0 \partial/\partial y^0 + y^1 \partial/\partial y^1 + y^2 \partial/\partial y^2$$

vanish. The properties (2) and (3) above correspond to these two conditions.

Hence we associated the form $\Omega(E)$ on V to the system E on P_2, and conversely the form $\Omega(E)$ determines the system E. When one performs a change of homogeneous coordinates on P_2, to treat the form $\Omega(E)$ is much easier than to struggle with the system E itself.

<u>The form $\Omega(\lambda)$ of Appell's equation $F_1(\lambda)$</u>: In § 10.5, the coefficients p_{ij}^k ($i,j,k = 1,2$) of Appell's equation $F_1(a,b,b',c)$ in canonical form E are given. If we put $a = \lambda_4$, $b = 1-\lambda_1$, $b' = 1-\lambda_2$ and $c = \lambda_3+\lambda_4$ ($\lambda_0+\lambda_1+\lambda_2+\lambda_3+\lambda_4 = 3$ as in § 10.5), then the coefficients $Q_{ijk}(y)$ ($i,j,k = 1,2,3$) of the corresponding form $\Omega(E)$, which we shall call $\Omega(\lambda)$, are given as follows. To make the expressions compatible with the indices of the λ's, we index the system of y-coordinates by y^1, y^2 and y^3 which are related to (x^1,x^2) through $x^1 = y^1/y^3$ and $x^2 = y^2/y^3$.

$$Q_{iii} = (1-\lambda_i)(y^j)^2(y^k)^2(y^j-y^k)^2/F,$$

$$Q_{iij} = (y^k)^2 y^j (y^j-y^k)\{(\lambda_0+\lambda_i-2)(y^k-y^i)(y^i-y^j)$$

$$+ (2-\lambda_i-\lambda_k)y^i(y^i-y^j) + (1-2\lambda_i+\lambda_j)y^i(y^k-y^i)\}/3F$$

for $\{i,j,k\} = \{1,2,3\}$, where

$$F = y^1 y^2 y^3 (y^1-y^2)(y^2-y^3)(y^3-y^1)$$

and Q_{123} is given by $y^1 Q_{123} + y^2 Q_{223} + y^3 Q_{332} = 0$.

§ 12.2 Integrability Conditions

We shall express the integrability condition on E in terms of the form ω.

Lemma: A system E in canonical form is integrable if and only if

$$p_{ij}^k = p_{ji}^k \qquad\qquad i,j,k = 1,2$$

$$p_{ij}^0 = -(p_{ij}^k)_k + p_{im}^k p_{jk}^m \qquad\qquad i,j = 1,2$$

and the expressions

$$(p_{ij}^0)_k + p_{ij}^m p_{km}^0 \qquad\qquad i,j,k = 1,2$$

are symmetric with respect to (i,j,k).

Proof: The argument in § 7.4 implies that the last expression above is symmetric with respect to (i,j,k) and that

$$p_{11}^0 = (p_{21}^2)_1 - (p_{11}^2)_2 + p_{21}^m p_{1m}^2 - p_{11}^m p_{2m}^2$$

$$p_{12}^0 = (p_{22}^2)_1 - (p_{12}^2)_2 + p_{22}^m p_{1m}^2 - p_{12}^m p_{2m}^2$$

$$p_{22}^0 = (p_{12}^1)_2 - (p_{22}^1)_1 + p_{12}^m p_{2m}^1 - p_{22}^m p_{1m}^1 .$$

By using the relation $p_{im}^m = 0$, we obtain the desired expressions for the p_{ij}^0's.

Let us define the differentiation d of a tensor

$$T = T_{i_1 \ldots i_m} dx^{i_1} \otimes \ldots \otimes dx^{i_m}$$

by

$$dT = (T_{i_1 \ldots i_m})_k dx^{i_1} \otimes \ldots \otimes dx^{i_m} \otimes dx^k .$$

For a tensor

$$S = S_{i_1 \ldots i_m j_1 \ldots j_n} dx^{i_1} \otimes \ldots \otimes dx^{i_m} \otimes dx^{j_1} \otimes \ldots \otimes dx^{j_n}$$

which is symmetric with respect to the i's and the j's, we define the <u>anti-symmetrizer</u> A by

$$AS = S_{i_1 \ldots i_m j_1 \ldots j_n} dx^{i_2} \otimes \ldots \otimes dx^{i_m} \otimes dx^{j_2} \otimes \ldots$$
$$\ldots \otimes dx^{j_n} (dx^{i_1} \wedge dx^{j_1}) .$$

We make the following calculation.

$$d\omega = (p_{ijk})_m dx^i \otimes dx^j \otimes dx^k \otimes dx^m (dx^1 \wedge dx^2)^{-1}$$

$$A d\omega = A(d\omega) = (p_{ijk})_m dx^i \otimes dx^j \frac{dx^k \wedge dx^m}{dx^1 \wedge dx^2}$$

$$\omega \otimes \omega = p_{ijk}\, p_{abc}\, dx^i \otimes dx^j \otimes dx^k$$
$$\otimes\, dx^a \otimes dx^b \otimes dx^c (dx^1 \wedge dx^2)^{-2}$$
$$A^2(\omega \otimes \omega) = A(A(\omega \otimes \omega))$$
$$= p_{ijk}\, p_{abc}\, dx^i \otimes dx^a\, \frac{dx^j \wedge dx^b}{dx^1 \wedge dx^2}\, \frac{dx^k \wedge dx^c}{dx^1 \wedge dx^2}$$

Put
$$\eta = A d\omega - A^2(\omega \otimes \omega)$$

then we have

$$\eta = \{(p_{ijk})_m\, \frac{dx^k \wedge dx^m}{dx^1 \wedge dx^2}$$
$$-\, p_{ijk}\, p_{abc}\, \frac{dx^j \wedge dx^b}{dx^1 \wedge dx^2}\, \frac{dx^k \wedge dx^c}{dx^1 \wedge dx^2}\,\} \, dx^i \otimes dx^j$$
$$= -\, \{(p_{ij}^k)_k + p_{im}^k\, p_{jk}^m\,\} \, dx^i \otimes dx^j$$
$$= p_{ij}^0\, dx^i \otimes dx^j.$$

Since we have

$$d\eta = (p_{ij}^0)_k dx^i \otimes dx^j \otimes dx^k,$$
$$\omega \otimes \eta = p_{ijk}\, p_{ab}^0\, dx^i \otimes dx^j \otimes dx^k$$
$$\otimes\, dx^a \otimes dx^b\, (dx^1 \wedge dx^2)^{-1},$$
$$A(\omega \otimes \eta) = p_{ija}\, p_{kb}^0\, dx^i \otimes dx^j \otimes dx^k\, \frac{dx^a \wedge dx^b}{dx^1 \wedge dx^2}$$
$$= -\, p_{ij}^b\, p_{kb}^0\, dx^i \otimes dx^j \otimes dx^k,$$

we conclude, by the previous lemma, that the system E is integrable if and only if the 3-tensor

$$d\eta - A(\omega \otimes \eta)$$

is symmetric. Thus we have expressed the integrability condition of E in a way which is free of coordinates. Since it is easy to check that $A(\omega \otimes A^2(\omega \otimes \omega))$ is symmetric, we have

<u>Lemma</u>: The system E is integrable if and only if the 3-tensor

$$d(Ad\omega) - d(A^2(\omega \otimes \omega)) - A(\omega \otimes Ad\omega)$$

is symmetric.

§ 12.3 The Restricted Riemann Problem

We consider the following restricted case of the Riemann problem.

<u>Problem</u>: Given an arrangement A in P_2 which consists of k distinct lines $L_1,\ldots,L_k$ and an complex number s_j on each L_j. Find a Fuchsian integrable system E on P_2 which has ramifying singularities along A with exponent s_j (§ 8.5) on each L_j.

We look for a rational tensor form

$$\Omega = Q_{abc}(y) dy^a \otimes dy^b \otimes dy^c \otimes \gamma(y)^{-1}$$

with the properties (1),(2),(3) and the following condition (4) which comes from the requirement of the local data and the proposition in § 8.5. Let l_j be a linear form defining A_j and put

$$L(y) = l_1 \cdots l_k.$$

Denote by $R_{abc}(y)$ the homogeneous polynomial $L(y)Q_{abc}(y)$ of degree k.

(4) Fix a value of j and let the coordinates $y = (y^0, y^1, y^2)$ be chosen such that the line A_j is defined by $y^1 = 0$, then

$$R_{122}(y), \quad R_{222}(y)/y^1, \quad R_{001}(y), \quad R_{000}(y)/y^1$$

$$R_{112}(y) - \frac{s_j - 1}{3} L(y) y^0 / y^1, \quad R_{011} + \frac{s_j - 1}{3} L(y) y^2 / y^1$$

are divisible by y^1.

For a given arrangement A, we can try to compute such a form Ω. The conditions (1),...,(4) are linear conditions on the coefficients of the polynomials $R_{abc}(y)$. We deliberately forget, for a moment, the integrability condition (which gives non-linear conditions on the coefficients) of the corresponding differential equation.

Remark: If A is the complete quadrilateral, then, since A consists of only six lines, the computation is simple and we can find the necessary and sufficient condition on the s_j's for Ω to exist and calculate the form Ω which turns out to be the form $\Omega(\lambda)$ corresponding to Appell's equation $F_1(\lambda)$ for some value of λ.

Proof: We put

$$L(y) = y^1 y^2 y^3 (y^1 - y^2)(y^2 - y^3)(y^3 - y^1)$$

and $\quad L(y) \Omega \gamma(Y) = R_{abc} \, dy^a \otimes dy^b \otimes dy^c$

where $R_{abc} = R_{bac} = R_{acb}$ (a,b,c = 1,2,3) are homogeneous polynomials of degree six. We use the notation $f(y)|g(y)$ to indicate that polynomial $f(y)$ is a factor of $g(y)$. Condition (4) along $\{y^a = 0\}$ ($a = 1,2,3$) can then be written

$$y^c | R_{bbc}, \quad (y^a)^2 (y^b)^2 | R_{ccc}$$
$$y^b | \{R_{bbc} - r_{bbc} (y^a)^2 y^c (y^a - y^b)(y^b - y^c)(y^c - y^a)\}$$

for $a \neq b \neq c \neq a$, where $r_{bbc} = r_{bba}$ are constants. Furthermore the equalities $\Sigma y^c R_{bbc} = 0$ of (3) implies that $(y^a)^2 y^c | R_{bbc}$ ($a \neq b \neq c \neq a$). Apply (4) to new coordinates $\underline{y}^1 = y^1 - y^0$, $\underline{y}^2 = y^2$, $\underline{y}^3 = y^3$. Then we have $(y^a)^2 (y^c - y^a) y^c | R_{bbc}$, $(y^a)^2 (y^a - y^b) y^b | R_{ccc}$ and so

$$R_{ccc} = r_{ccc} (y^a)^2 (y^a - y^b)^2 (y^b)^2$$
$$R_{bbc} = (y^a)^2 (y^c - y^a) y^c \{ r_{bbc} (y^a - y^b)(y^b - y^c)$$
$$+ r'_{bbc} y^b (y^b - y^c) + r''_{bbc} y^b (y^a - y^b) \}$$

for $a \neq b \neq c \neq a$, where r' and r'' are constants. The condition (3) requires that $\Sigma y^c R_{abc} = 0$ ($a,b = 1,2,3$) and so gives a system of linear homogeneous equations for r_{aaa}, $r_{aab} = r_{aac}$, r'_{aab} and r''_{aab} ($a \neq b \neq c \neq a$). It is now easy to show that this system has a 4-dimensional solution space. Indeed R_{123}, R_{112}, R_{221}, R_{113} and R_{223} are expressed linearly in terms of R_{333}, R_{111}, R_{222}, R_{331} and R_{332}. We have relations

$$r''_{331} = r_{333} - r'_{332}, \qquad r''_{332} = r_{333} - r'_{331},$$
$$r_{111} = -r_{333} + 3r'_{332}, \qquad r_{222} = -r_{333} + 3r'_{331},$$

which show that all the r, r' and r" are linear combinations of r_{333}, r'_{331}, r'_{332} and $r_{331} = r_{332}$. We have only to put

$$r_{333} = 3(1-\lambda_3), \qquad r_{331} = r_{332} = \lambda_0 + \lambda_3 - 2,$$
$$r'_{331} = 2 - \lambda_3 - \lambda_2, \qquad r'_{332} = 2 - \lambda_3 - \lambda_1$$

to obtain the expression of $\Omega(\lambda)$ in § 12.1. This completes the proof.

If A is the arrangement $A_m^0(3)$ or $A_m^3(3)$, then the rational form Ω reduces to the form $\Omega(\lambda)$ for Appell's system for some λ. More precisely we can prove

<u>Lemma</u> ([Yos 2]): Any rational form Ω with the properties (1),...,(4) for the arrangement

$$y^0 y^1 y^2 ((y^0)^m - (y^1)^m)((y^1)^m - (y^2)^m)((y^2)^m - (y^0)^m) = 0$$

is the pullback by the map $V \to \mathbb{C}^3$ given by

$$(y^0, y^1, y^2) \mapsto (w^0, w^1, w^2) = ((y^0)^m, (y^1)^m, (y^2)^m)$$

of the form $\Omega(\lambda)$ for some λ.

For each of the other arrangements A defined by 3-dimensional exceptional unitary reflection groups G, we assume that the equation E is invariant under the group G, that is, we look for a form Ω which is G-invariant.

§ 12.4 G-Invariant System of Differential Equations I

Let V be a 3-dimensional unitary space and $G \subset U(V)$ be a unitary reflection group which is either Icosahedral, Klein or

Valentiner (cf.§ 11.3, § 12.4). Let $l_1,\ldots,l_k$ be linear forms which define the mirrors $A_1,\ldots,A_k$ in V (as well as lines $L_1,\ldots,L_k$ in P_2) of the reflections in G, and let $L = L(y) = l_1\ldots l_k$. Finally let

$$\Omega = \frac{R_{abc}(y)}{L(y)} dy^a \otimes dy^b \otimes dy^c \otimes \gamma(y)^{-1}$$

be a G-invariant rational form on V with properties (1),...,(4), The group G acts transitively on the set of planes $A_1,\ldots,A_k$. Thus we can put

$$s = s_1 = \ldots = s_k \quad (\neq 1)$$

Lemma: The symmetric polynomial form on V given by

$$\Omega' = R_{abc}(y) \, dy^a \odot dy^b \odot dy^c$$

is G-invariant.

Proof: From the table in § 11.4, we see that the group G is generated by reflections of order 2. Any such reflection in G changes the sign of the polynomial $L(y)$ and of the form $\gamma(y)$. Therefore $L(y)\gamma(y)$ is G-invariant and so is $\Omega' = \Omega L(y)\gamma(y)$.

Let $\odot$ denote the symmetric tensor product. That is, for example,

$$dy^n \odot dy^m \odot dy^l = (3!)^{-1}(dy^n \otimes dy^m \otimes dy^l + dy^n \otimes dy^l \otimes dy^m + dy^m \otimes dy^n \otimes dy^l + \ldots)$$

Lemma: Let A_a, A_b and A_c be three mutually orthogonal planes in A. Then we have

$$\Omega' = (s-1)J(a,b,c)^{-1} \frac{L}{l_b} (1_a dl_c - 1_c dl_a) \odot (dl_b)^2$$

modulo l_b, where $J(a,b,c) = \det(\partial(1_a,1_b,1_c)/\partial(y^0,y^1,y^2))$.

<u>Proof</u>: Choose a system of coordinates $\underline{y} = (\underline{y}^0,\underline{y}^1,\underline{y}^2)$ such that $\underline{y}^0 = 1_a$, $\underline{y}^1 = 1_b$ and $\underline{y}^2 = 1_c$. Let $\underline{R}_{abc}(\underline{y})$ be the coefficients of the tensor $\mathfrak{L}(y(\underline{y}))\gamma(\underline{y})^{-1}$ with respect to the coordinate $\underline{y}$. Since $\gamma(\underline{y})\gamma(y)^{-1}$ is a constant which is equal to $\det(\partial(1_a,1_b,1_c)/\partial(y^0,y^1,y^2))$, we have

$$\Omega' = J(a,b,c)^{-1} \underline{R}_{mn1}(\underline{y}) d\underline{y}^m d\underline{y}^n d\underline{y}^1.$$

The invariance of Ω' under G implies in particular that it is invariant under the reflection with mirror A_b. This reflection is represented, in the coordinate $\underline{y}$, by

$$(\underline{y}^0,\underline{y}^1,\underline{y}^2) \to (\underline{y}^0, -\underline{y}^1,\underline{y}^2).$$

Therefore the above expression for Ω' implies that

$$\underline{R}_{mn1}(\underline{y}) = 0 \quad \text{modulo } \underline{y}^1$$

if $\{m,n,1\} = \{1,1,1\}$ or $\{1,p,q\}$ (where $p,q \neq 1$). On the other hand property (4) tells us that $\underline{R}_{222}(\underline{y}) \equiv 0$ modulo $\underline{y}^1$. Moreover property (3) with respect to the coordinate $\underline{y}$ implies that, for $m,n = 0,1,2$ we have

$$\underline{y}^0 \underline{R}_{mn0}(\underline{y}) + \underline{y}^2 \underline{R}_{mn2}(\underline{y}) = 0 \quad \text{modulo } \underline{y}^1.$$

By these equalities, we know that all the $\underline{R}_{mn\ell}(\underline{y})$ except for $\underline{R}_{110}(\underline{y})$ and $\underline{R}_{112}(\underline{y})$ are zero modulo $\underline{y}^1$. Since the remaining two are related by the identity above, we have

$$\underline{R}_{110}(\underline{y}) \equiv -3^{-1}(s-1)L(y(\underline{y}))\underline{y}^2/\underline{y}^1 \quad \text{modulo} \quad \underline{y}^1.$$

This completes the proof.

<u>Lemma</u>: For each plane A_b in A, we can choose two planes $A_{b'}$ and $A_{b''}$ in A in such a way that the three planes A_b, $A_{b'}$ and $A_{b''}$ are orthogonal to each other and that the set of triples $\{A_b, A_{b'}, A_{b''}\}$ $(b = 1,\ldots,k)$ is G-invariant.

<u>Proof</u>: If G is the icosahedral group, then the picture of the arrangement in § 10.3 tells us that, for each plane, there are exactly two planes in A such that these three planes form an orthogonal triplet. Let G be the Klein or Valentiner group and let S be the set of triples of planes in A which are mutually perpendicular. For each plane A_b, there are four planes A_a, A_c, A_d and A_e in A such that $\{A_a, A_b, A_c\}$, $\{A_d, A_b, A_e\} \in S$ (cf. [S-T]). In particular we have $\#S = 2k/3$. The group G acts on S; Let H be the isotropy subgroup of G at $\{A_a, A_b, A_c\}$. The group H contains the reflection in the mirror A_a and the symmetric group S_3 which permutes the three planes. If we identify $U(1)$ with the centre of $U(3)$ we find that, for the Valentiner group, H also contains the cube roots of unity. Thus $|H| \geq 2^3 3!$ and so $|G/H| \leq 7$ for the Klein group; and $|H| \geq 3 \cdot 2^3 3!$ and so $|G/H| \leq 15$ for the Valentiner group. As G acts transitively on the planes of S, these inequalities are equalities. Now since $|S| = 14$ and 30 for the two groups, respectively, we conclude that S is the union of two G-orbits

$$S = G\{A_a, A_b, A_c\} \sqcup G\{A_d, A_b, A_e\}$$

This completes the proof.

Consider the following symmetric form Ω'':

$$(s-1) \sum_{b=1}^{k} J(b',b,b'')^{-1} \frac{L}{l_b}(l_{b'}dl_{b''} - l_{b''}dl_{b'}) \circ (dl_b)^{\circ 2}$$

Since the system $\{A_b, A_{b'}, A_{b''}\}$ $(b = 1,\ldots,k)$ is G-invariant and the group G is generated by reflections in the mirrors $A_1,\ldots A_k$, the form Ω'' is G-invariant. Thus $\Omega' - \Omega''$ is a G-invariant form such that

$$\Omega' - \Omega'' \equiv 0 \quad \text{modulo } l_b \quad (b = 1,\ldots,k).$$

Therefore $(\Omega'-\Omega'')/L(y)$ is an G-anti-invariant polynomial form (§ 11.1) of degree 3. This implies that $\Omega' = \Omega''$.

We have shown that if a G-invariant form Ω has the properties (1),...,(4) for the Icosahedral, Klein or Valentiner group G, then Ω has the form $\Omega(s)$:

$$(s-1) \sum_{b=1}^{k} J(b',b,b'')^{-1} \frac{l_{b'}dl_{b''} - l_{b''}dl_{b'}}{l_b} \circ (dl_b)^{\circ 2} \gamma(y)^{-1}$$

We check that $\Omega(s)$ has indeed the property (3). The interior product of $\Omega(s)$ with respect to the vector

$$y^i \partial/\partial y^i = l_{b'} \partial/\partial l_{b'} + l_b \partial/\partial l_b + l_{b''} \partial/\partial l_{b''}$$

is by definition equal to

$$(s-1)\Sigma_{b=1}^{k}J(b',b,b'')^{-1}\frac{1}{3l_b}\{(l_b,dl_{b''}-l_{b''}dl_{b'})\otimes dl_b\ l_b$$
$$+dl_b\otimes(l_b,dl_{b''}-l_{b''}dl_{b'})\ l_b$$
$$+dl_b\otimes dl_b(l_{b'},\ l_{b''}-l_{b''}\ l_{b'})\ \}$$
$$+\frac{2}{3}(s-1)\Sigma_{b=1}^{k}J(b',b,b'')^{-1}(l_b,dl_{b''}-l_{b''}dl_{b'})\odot dl_b$$

A cyclic change of indices b',b,b'' shows that this expression is zero.

We finally show that the corresponding differential equation is integrable. Let $\omega(s)$ be the form corresponding to the form $\Omega(s)$, obtained by putting $y^0=1$, $y^1=x^1$, $y^2=x^2$. For notational simplicity, we keep the same notation l_a for $l_a(1,x^1,x^2)$, and write

$$J(a,b)=\det\ (\partial(l_a,l_b)/\partial(x^1,x^2))$$

In the calculation below, we will calculate explicitly the tensor $d(Ad\omega(s))$ and see that it is symmetric. We have

$$d\omega(s)=(s-1)\Sigma_{b=1}^{k}\ J(b',b,b'')^{-1}$$
$$\times\{\frac{1}{l_b}(dl_{b''}\odot dl_b\odot dl_b\otimes dl_{b'}-dl_{b'}\odot dl_b\odot dl_b\otimes dl_{b''})$$
$$-\frac{1}{(l_b)^2}(l_b,dl_{b''}-l_{b''}dl_{b'})\odot dl_b\odot dl_b\otimes dl_b)\}\ (dx^1\wedge dx^2)^{-1}$$

Since we have in general

$$A(df_1\odot\ldots\odot df_n\otimes dg)=(n)^{-1}\Sigma_{j=n}^{n}df_1\odot\ldots\hat{j}\ldots\odot df_n(df_j\wedge dg)$$

(the symbol $\hat{j}$ means that df_j is omitted) we have

$$Ad\omega(s) = 3^{-1}(s-1) \Sigma_{b=1}^{k} J(b',b,b")^{-1}$$

$$\times \{(1_b)^{-1}(2dl_{b"}\circ dl_b J(b,b') + dl_b \circ dl_b J(b",b')$$

$$- 2dl_{b'}\circ dl_b J(b,b") - dl_b \circ dl_b J(b',b"))$$

$$- (1_b)^{-2}(1_b,dl_b\circ dl_b J(b",b) - 1_{b"}dl_b \circ dl_b J(b',b))\}$$

$$d(Ad\omega(s)) = 3^{-1}(s-1) \Sigma_{b=1}^{k} J(b',b,b")^{-1}$$

$$\times \{(1_b)^{-2}(-2dl_{b"}\circ dl_b \otimes dl_b J(b,b')$$

$$- dl_b \circ dl_b \otimes dl_b J(b",b') + 2dl_{b'}\circ dl_b \otimes dl_b J(b,b")$$

$$+ dl_b \circ dl_b \otimes dl_b J(b',b")) - dl_b \circ dl_b \otimes dl_b J(b",b)$$

$$+ dl_b \circ dl_b \otimes dl_{b"} J(b',b))$$

$$+ 2(1_b)^{-3}(1_b,dl_b\circ dl_b \otimes dl_b J(b",b)$$

$$- 1_{b"}dl_b \circ dl_b \otimes dl_b J(b',b))\}$$

$$= (s-1) \Sigma_{b=1}^{k} J(b',b,b")^{-1}$$

$$\times \{(1_b)^{-2}(dl_{b"} \circ dl_b \circ dl_b J(b',b)$$

$$+ dl_{b'} \circ dl_b \circ dl_b J(b,b")$$

$$+ (2/3)(1_b)^{-3}(1_b,J(b",b) - 1_{b"}J(b',b))(dl_b)^{\circ 3}\}$$

As mentioned above, this is seen to be symmetric. Hence we only have to show that the tensor

$$I(s) = d(A^2(\omega(s) \otimes \omega(s))) + A(\omega(s) \otimes Ad\omega(s))$$

is symmetric. Since we have

$$I(s) = (s-1)^2 I(0)$$

it suffices to show that $I(s)$ is symmetric for a special value of $s \neq 1$. We have already shown that for some weight $1/n$ (given in the table in the end of § 10.3), the orbifold (X,S,b) ($X-S = P_2-A$) constructed in § 10.3 is uniformized by the ball. So there is a unique uniformizing equation in canonical form on P_2 (§ 10,4), which is invariant under G. The corresponding form Ω is G-invariant and has the prperties (1),...,(4). Therefore it coincides with $\Omega(1/n)$. Since the uniformizing equation is certainly integrable, $I(1/n)$ is symmetric. We have now completed the proof of the following result.

<u>Theorem</u>: Let $A \subset P_2$ be a line arrangement defined by either Icosahedral, Klein or Valentiner group G. For each complex number s ($\neq 1$), there is a G-invariant Fuchsian differential equation on P_2 which has ramifying singularities of exponent s along A. The corresponding form $\Omega(s)$ on V is given by

$$(s-1) \sum_{b=1}^{k} J(b',b,b'')^{-1} \frac{l_b \cdot dl_{b''} - l_{b''} \cdot dl_{b'}}{l_b} \odot (dl_b)^{\odot 2} \gamma(y)^{-1}$$

If n is one of the weights given in § 10.3 then the equation for s = 1/n is the uniformizing equation of the corresponding orbifold given in § 10.3.

§ 12.5 G-Invariant System of Differential Equations II

Let G be the Hesse or extended Hesse group (cf. § 10.3 and § 11.4), and let G' be the projectivized group, which is the same for both groups. For the arrangement A defined by G, we construct G-invariant differential equations with ramifying singularities along A.

We study the arrangement in detail and find geometric relations with the complete quadrilateral, which imply the existence of a deep relation between the equation in question and Appell's equation F_1.

Let $x = (x^0, x^1, x^2)$ be a system of homogeneous coordinates on $X = \mathbb{P}_2$. Following [S-T], we put

$$I_6(x) = ((x^0)^3 + (x^1)^3 + (x^2)^3)^2$$
$$\qquad - 12((x^0 x^1)^3 + (x^1 x^2)^3 + (x^2 x^0)^3),$$
$$I_9(x) = ((x^0)^3 - (x^1)^3)((x^1)^3 - (x^2)^3)((x^2)^3 - (x^3)^3)$$
$$I_{12}(x) = ((x^0)^3 + (x^1)^3 + (x^2)^3)(((x^0)^3 + (x^1)^3$$
$$\qquad + (x^2)^3))^3 + 216(x^0 x^1 x^2)^3),$$
$$I'_{12}(x) = x^0 x^1 x^2 \prod_{a,b=0}^{2}(\omega^a x^0 + \omega^b x^1 + x^2)$$

where $\omega = \exp(2\pi i/3)$. We have the relation

$$6912(I'_{12})^3 = (432\, I_9^2 - I_6^3 + 3 I_6\, I_{12})^2 - 4(I_{12})^3.$$

We define two kinds of arrangements on X as follows.

$$A'_X : I'_{12}(x) = 0$$
$$A''_X : I_9(x) = 0$$

The Hesse arrangement is A'_X, while the union of A'_X and A''_X is the extended Hesse arrangement.

Let S be the quotient variety of X by the group G' and $p: X \to S$ be the projection. The variety S is <u>the weighted projective space of type</u> $(2,3,4)$, that is,

$$S = \{(s_2, s_3, s_4) \in \mathbb{C}^3 - \{0\}\}/\sim$$

where $(s_2, s_3, s_4) \sim (\lambda^2 s_2, \lambda^3 s_3, \lambda^4 s_4)$ for $\lambda \in \mathbb{C}-\{0\}$. The map p is given by

(p) $\qquad s_2 = I_6(x), \qquad s_3 = I_9(x), \qquad s_4 = I'_{12}(x);$

its ramification locus on S is

$$A'_S := p(A'_X) \quad \text{and} \quad A''_S := p(A''_X)$$

with indices 3 and 2, respectively. We have

$$A'_S : \quad (432s_3^2 - s_2^3 + 3s_2s_4)^2 - 4s_4^3 = 0$$
$$A''_S : \quad s_3 = 0.$$

Let Y be another projective plane with a homogeneous coordinate system $y = (y^0, y^1, y^2)$. Putting $y^3 = -y^0 - y^1 - y^2$ we find that the symmetric group S_4 acts on Y by permutating the four coordinates. Let T be the quotient of Y by the group S_4 with projection $q : Y \to T$. The variety T is again the weighted projective space of type (2,3,4). In terms of the weighted homogeneous coordinates $t = (t_2, t_3, t_4)$ the map q is given by

$$\text{(q)} \quad \begin{aligned} t_2 &= \Sigma_{0 \le i < j \le 3} \, y^i y^j \\ t_3 &= - \Sigma_{0 \le i < j < k \le 3} \, y^i y^j y^k \\ t_4 &= y^0 y^1 y^2 y^3 \end{aligned}$$

The group S_4 has fixed points of multiplicity 2 on Y along the curves

$$A'_Y := \bigcup_{0 \le i < j \le 2} \{y^i - y^j = 0\}$$
$$\bigcup_{\{i,j,k\}=\{0,1,2\}} \{2y^i + y^j + y^k = 0\}$$
$$A''_Y := \bigcup_{0 \le i < j \le 2} \{y^i + y^j = 0\}$$

The branch locus of q on T is given by

$$A_T' := q(A_Y') \quad \text{and} \quad A_T'' := q(A_Y'')$$

By computing the discriminant of the polynomial of degree 4

$$x^4 + t_2 x^2 + t_3 x + t_4,$$

we know the defining equation of A_T':

$$(t_3^2 - (-2t_2/3)^3 + 3(-2t_2/3)(4t_4+t_2^2/3)/3)^2$$
$$- 4((4t_4 + t_2^2/3)/3)^3 = 0$$

Putting the expression $y^3 = -y^0 - y^1 - y^2$ into (q), we have

$$t_3 = (y^0 + y^1)(y^1 + y^2)(y^2 + y^0)$$

Thus we have the equation of A_T'' : $t_3 = 0$
Let ψ be the map $S \to T$ given by

$$(\psi) \quad s_2 = -2t_2/3, \quad s_3 = t_3/\sqrt{432}, \quad s_4 = 4t_4/3 + t_2^2/9$$

Then ψ gives an isomorphism $S \to T$ which induces the isomorphisms

$$A_S' \to A_T' \quad \text{and} \quad A_S'' \to A_T''.$$

Let Z be another projective plane with homogeneous coordinates $z = (z^0, z^1, z^2)$. Put

$$A_Z' : (z^0 - z^1)(z^1 - z^2)(z^2 - z^0) = 0$$
$$A_Z'' : z^0 z^1 z^2 = 0$$

and let r be the map $Y \to Z$ given by

(r) $\quad z^0 = (y^1 + y^2)^2, \quad z^1 = (y^2 + y^0)^2, \quad z^2 = (y^0 + y^1)^2$

The map r, which is the quotient map by $(\mathbb{Z}/2\mathbb{Z})^2$, ramifies along A_Z'' with index 2. We have

$$r(A_Y') = A_Z', \quad r(A_Y'') = A_Z''$$

Summing up, we have had the following diagram

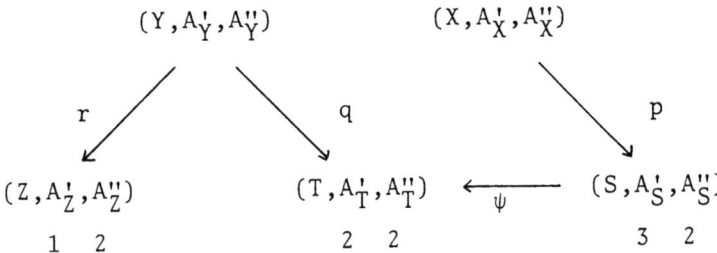

Here the numerals below the A's denote the branching indices of the corresponding coverings.

We are looking for a G-invariant differential equation E on X with ramifying singularities along $A = A_X' \cup A_X''$. If there is such E, then it is defined on S, so on T. Lifting the equation by q we have a system on Y. On the other hand the lemma in § 12.3 tells us that it is the lift by r of an Appell's equation $F_1(\lambda)$. Let the exponents of E along A_X' and A_X'' be s' and s", respectively. Since the exponents of the $F_1(\lambda)$ along A_Z' and A_Z'' are $2s'/3$ and $s"/2$, respectively, we have

$$\lambda_0 = 1/2 + 2s'/3 - s"/4$$
$$\lambda_1 = \lambda_2 = \lambda_3 = 1/2 + s"/4$$
and $\quad \lambda_4 = 1 - 2s'/3 - s"/2.$

Thus our system E is obtained from the $F_1(\lambda)$ by the algebraic transformation

$$x = p^{-1} \cdot \psi^{-1} \cdot q \cdot r^{-1}(z)$$

of the independent variables. Computing the transformations r, q, ψ and p we have

Theorem: For complex numbers s', s'', there is a G-invariant Fuchsian differential equation on $\mathbb{P}_2$ which has ramifying singularities along A_X' and A_X'' of exponents s' and s'', respectively. An explicit form for the system is given, in inhomogeneous coordinates $x = x^1/x^0$, $y = x^2/x^0$, as follows

$$p_{11}^1 = 3t'/x + 81t'x^2y^3(2-x^3-y^3)/W + 3t''x^2(y^3-1)(1+x^3-2y^3)/2H$$

$$p_{11}^2 = 81t'xy(1+x^3-y^6-x^3y^3)/W - 9t''xy(y^3-1)^2/2H$$

$$p_{22}^2 = 3t'/y + 81t'y^2x^3(2-y^3-x^3)/W - 3t''y^2(x^3-1)(1+y^3-2x^3)/2H$$

$$p_{22}^1 = 81t'yx(1+y^3-x^6-y^3x^3)/W + 9t''yx(x^3-1)^2/2H$$

where

$$W = \prod_{a,b=0}^{2}(\omega^a x + \omega^b y + 1) = (x^3 + y^3 + 1)^3 - 27x^3y^3$$

$$H = (x^3 - 1)(y^3 - 1)(x^3 - y^3)$$

$$t' = (s' - 1)/3, \quad t'' = (s'' - 1)/3.$$

If n' and n'' are weights of Hesse lines and the additional lines, respectively, given in § 10.3, then the equation for $s' = 1/n'$ and $s'' = 1/n''$ is the uniformizing equation of the corresponding orbifold given in § 10.3.

References

[Aom 1] K. Aomoto, Un théorème du type Matsushima-Murakami concernant l'intégrale des fonctions multiformes, J. Math. Pures Appl., 52 (1973), 1-11.

[Aom 2] K. Aomoto, Les équations aux différences linéaires et les intégrales des fonctions multiformes, J. Fac. Sci. Univ. Tokyo, Sec. IA, 22 (1975), 271-297 and: Une correction et un complément a l'article "Les équations aux différences linéaires et les intégrales des fonctions multiformes", ibid., 26 (1979), 519-523.

[Aom 3] K. Aomoto, On vanishing of cohomology attached to certain many valued meromorphic functions, J. Math. Soc. Japan, 27 (1975), 248-255.

[Aom 4] K. Aomoto, On the structure of integrals of power product of linear functions, Sci. Papers College Gen. Ed. Univ. Tokyo, 27 (1977), 49-61.

[A-K] P. Appell & J. Kampé de Feriet, Fonctions Hypergéometriques et Hyperspheriques-Polynomes d'Hermite, Gauthier-Villars, Paris, (1926).

[Arm] M.A. Armstrong, On the fundamental group of an orbit space, Proc. Cambridge Philos. Soc., 61 (1965), 639-646.

[Bie] L. Bieberbach, Theorie der gewöhnlichen Differentialgleichungen, Springer-Verlag, 1953.

[BHH]   G. Barthel, F. Hirzebruch & T. Höfer,Geraden
         -konfiguration und Algebraische Flächen, Vieweg
         Aspekte der Mathematik

[Bl]   A. Boulanger, Contribution à l'étude des équations
         différentielles linéaires homogènes intégrables
         algébriquement, J. Ec. Pol., 4 (1898), 1-122.

[Br]   N. Bourbaki, Groupes et Algebres de Lie, Ch. 4, 5 et
         6, Hermann, Paris, (1968).

[BBRWZ] H. Brown, R. Bulow, J. Neubuser, H. Wondratschek &
         H. Zassenhaus, Crystallographic groups of four-
         dimensional space, John Wiley & Sons, (1978).

[Che]   C. Chevalley, Invariants of finite groups generated
         by reflections, Amer. J. Math., 77 (1955), 778-782.

[Del]   P. Deligne, Equations differentielles a points
         singuliers réguliers, Lecture Notes in Math., 163,
         Berlin-Heidelberg-New York: Springer-Verlag, (1970).

[D-M]   P. Deligne & G.D. Mostow, Monodromy of
         hypergeometric functions and non-lattice integral
         monodromy, I.H.E.S. Publication Math. No. 63 (1986) 5-
         106.

[Erd]   A. Erdelyi, Higher transcendental functions, McGraw-
         Hill, 1 (1953).

[Fox 1] R.H. Fox, On Fenchel's conjecture about F-groups,
         Matematisk Tidsskrift B, (1952), 61-65.

[Fox 2] R.H. Fox, Covering spaces with singularities,
         Algebraic geometry and topology, 243-257.

[G-O]　　R. Gérard & K. Okamoto (editors), Equations différentielles dans le champ complexe (colloque franco-japonais 1985) Hermann (1987).

[Gol]　　Private communication with W.M. Goldman, (1984).

[Hat]　　A. Hattori, Topology of $\mathbb{C}^n$ minus a finite number of affine hyperplanes in general position, J.Fac.Sci. Univ. of Tokyo, 22 (1975), 205-219.

[H-K]　　A. Hattori & T. Kimura, On the Euler integral representations of hypergeometric functions in several variables, J. Math. Soc. Japan, 26 (1974), 1-16.

[Hir 1]　F. Hirzebruch, Arrangements of lines and algebraic surfaces, In: Arithmetic and geometry, Vol.II, Progress in Math., 36, 113-140, Boston, Basel, Stuttgart: Birkhäuser, (1983).

[Hir 2]　F. Hirzebruch, Chern numbers of algebraic surfaces, Math. Ann., 266 (1984), 351-356.

[Hir 3]　F. Hirzebruch, Algebraic surfaces with extreme Chern numbers, Preprint MPI Bonn(1984).

[Hir 4]　F. Hirzebruch, The ring of Hilbert modular forms for real quadratic fields of small discriminant. Lec. Note in Math. Springer 627 (1977) 287-323.

[Hir 5]　F. Hirzebruch, Automorphe Formen und der Satz von Riemann-Roch, Symp. Int. Top. Alg. 1956, México, Univ. México (1958) 129-144.

[Höf]　　Th. Höfer, Ballquotienten als verzweigte Überlegungen der projectiven Ebene, Dissertation, Bonn, (1985).

[Hol 1] R.-P. Holzapfel, Invariants of arithmetic ball quotient surfaces, Math. Nachr., 103 (1981), 117-153.

[Hol 2] R.-P. Holzapfel, Arithmetic curves on ball quotient surfaces, Ann. Glob. Analysis and Geometry 1, No.2, (1983), 21-90.

[Huk] M. Hukuhara, Ordinary differential equations (in Japanese) Iwanami Zensho 116 (1980)

[Hun] B. Hunt, Coverings and ball quatients with special emphasis on the 3-dimensional case, Bonner Math. Schriften, Nr.174 (1986).

[Inu] T. Inui, Special functions (in Japanese) Iwanami(1962)

[Iwa] M. Iwano, Ordinary differential equations (in Japanese) Lec. Note, Tokyo Metropolitan Univ. (1985)

[J-M] D. Johnson & J.J. Millson, Deformation spaces associated to compact hyperbolic manifolds, to appear in "Discrete Groups in Geometry and Analysis"

[Kan] J. Kaneko, Monodromy group of Appell's System (F_4), Tokyo J. of Math., 4 (1981), 35-54.

[K-T-Y] J. Kaneko, S. Tokunaga & M. Yoshida, , Complex crystallographic groups II, J. Math. Soc. Japan, 34-4 (1982), 594-605.

[Kat 1] M. Kato, On the existence of finite principal uniformizations of $\mathbb{C}P^2$ along weighted line configurations, Mem. Fac. Sci. Kyushu Univ. Ser. A, 38 (1984), 127-131.

[Kat 2] M. Kato, On uniformization of orbifolds (preprint).

[Kim 1] T. Kimura, Hypergeometric functions of two variables, Lecture Notes Univ. of Minnesota, (1971-1972).

[Kim 2] T. Kimura, On the isomonodromic deformation for linear ordinary differential equations of the second order, Proc. Japan Acad.,57,(1981), 285-290.

[K-O 1] H. Kimura & K. Okamoto, On the isomonodromic deformation of linear ordinary differential equations of higher order, Funkcial. Ekvac., 26 (1983), 37-50.

[K-O 2] H. Kimura & K. Okamoto, On the polynomial Hamiltonian structure of the Garnier system, J. Math. Pures et Appl. (1984), 129-146.

[Ki-No] M. Kita & M. Noumi, On the structure of cohomology groups attached to the integral of certain many-valued analytic functions, Jap. J. of Math. 9 (1983), 113-157.

[K-N] R. Kobayashi & I. Naruki, Holomorphic conformal structures and uniformization of complex surfaces (preprint).

[Kg] S. Kitagawa, On the classification of parabolic reflection groups, to appear in J. of Math, Soc. of Japan.

[KoR] R. Kobayashi, Einstein-Kähler V-metrics on open Satake V-surfaces with isolated quotient singularities, Math. Ann., 272 (1985), 385-398.

[Ko-O] S. Kobayashi & T. Ochiai, Holomorphic projective structures of compact complex surfaces, Math. Ann. 249 (1980), 75-94.

[Mil]   W. Miller, Jr., Lie theory and the Lauricella functions F_D, J. Math. Phys. 13 (1972), 1393-1399.

[Miy 1] Y. Miyaoka, On the Chern numbers of surfaces of general type, Inv. Math. 42 (1977), 225-237.

[Miy 2] Y. Miyaoka, On algebraic surfaces with positive index, Progress in Mathematics (Birkhäuser Bostn) Vol. 39 (1983), 281-301.

[Mos 1] G.D. Mostow, Strong rigidity of locally symmetric spaces, Ann. Math. Studies, 78, Princeton Univ. Press, (1973).

[Mos 2] G.D. Mostow, Existence of nonarithmetic monodromy groups, Proc. Nat. Acad. Sci. USA, 78 (1981), 5948-5950.

[Mum]   D. Mumford, The topology of normal singularities of an algebraic surface and a criterion for simplicity, Inst. Hautes Etudes Sci. Publ. Math., 9 (1961), 229-246.

[Nak]   E. Nakagiri, Monodromy representations of hypergeometric differential equations of two variables, Master Theses, Kobe Univ. (1979).

[Nam 1] M. Namba, On branched coverings of projective manifolds, Proc. Jap. Acad., 61 (1985), 121-124.

[Nam 2] M. Namba, Lectures on branched coverings and algebraic functions, (1985).

[Oda]   T. Oda, On Schwarzian derivatives in several variables (in Japanese), Kokyuroku of R.I.M., Kyoto Univ., 226 (1974).

[Oku]   K. Okubo, On the group of Fuchsian equations, Progress report for Grant-in-Aid for Scientific Reseach (Ministry of Education, Japan) (1981)

[Oht 1]   M. Ohtsuki, A residue formula for Chern classes associated with logarithmic connections, Tokyo J. of Math. 5 (1982) 13-21.

[Oht 2]   M. Ohtsuki, On the number of apparent singularities of a linear differential equations, Tokyo J. of Math. 5 (1982) 23-29.

[Oka 1]   K. Okamoto, Introduction to the Painlevé equations (in Japanese), Sophia Kokyuroku in Math., 19 (1985).

[Oka 2]   K. Okamoto, Sur les échelle associées aux fonctions spéciales et l'équation de Toda, preprint.

[Oka 3]   K.Okamoto, Isomonodromic deformation and Painlevé equations and the Garnier system, J. Fac. Sci. Tokyo Univ., 33 (1986).

[O-K]   K. Okamoto & H. Kimura, On particular solutions of the Garnier systems and the hypergeometric functions of several variables. Quart. J. Math. Oxford(2), 37 (1986), 61-80.

[Pia]   I.I. Piatetskii-Shapiro, Automorphic Functions and the Geometry of Classical Domains, Gordon and Breach, New York, (1969).

[Pic 1]   E. Picard, Sur les fonctions de deux variables indépendantes analogues aux fonctions modulaires, Acta Math., 2 (1883), 114-126.

[Pic 2]   E. Picard, Sur les fonctions hyperfuchsiennes provenant des séries hypergeométriques de deux variables, Ann. Sci. Ecole Norm. Sup. III, 2 (1885), 357-384.

[Röh] H. Röhrl, Das Riemann-Hilbertsche Problem der Theorie der linearen Differentialgleichungen, Math. Ann., 133(1957), 1-25.

[Sak] F. Sakai, Defect relations and ramifications, Proc. Japan Acad., 50 (1974), 723-728.

[Ssi] T. Sasai, On a monodromy group and irreducible conditions of a fourth order Fuchsian differential system of Okubo type, J. für die reine und angewandte Math. 299/300 (1978), 38-50.

[Ssk] T. Sasaki, On the finiteness of the monodromy group of the system of hypergeometric differential equations (F_D), J. Fac. Sci. Univ. of Tokyo, 24 (1977), 565-573.

[Sa-Ta] T. Sasaki & K. Takano, On Appell's systems of hypergeometric differential equations, Recent developments in several complex variables, Princeton Univ. Press (1981).

[S-Y] T. Sasaki & M. Yoshida, Differential equations in two variables of rank 4, MPI preprint (1986).

[Sat] M. Sato, Singular orbits in prehomogeneous vector spaces, Lec. Note at Univ. of Tokyo (1972)

[SKK] M. Sato, T. Kawai & M.Kashiwara, Micro-functions and pseudo-differential equations, Lec. Notes in Math. 287, Springer (1971).

[Sel] A. Selberg, On discontinuous groups in higher-dimensional symmetric spaces, Contributions to function theory, Tata Inst. Bonbay (1960), 147-164.

[Sch]   H.A. Schwarz, Ueber diejenigen Fälle, in welchen die Gaussische hypergeometrische Funktion algebraische Funktion ihres vierten Elementes darstellt, J. Reine Angew. Math. (J. de Crelle), 54 (1873), 292-335.

[S-T]   G.C. Shephard & J.A. Todd, Finite unitary reflection groups, Canad. J. Math., 6 (1954), 274-304.

[Shg 1] H. Shiga, One attempt to the K3 modular function I, II, Ann. Scuola Norm. Pisa Serie IV-Vol.VI (1979), 609-635; Serie IV-Vol.VIII (1981), 157-182.

[Shg 2] H. Shiga, On the representation of Picard modular function by theta constants I, II (preprints).

[Shm]   G. Shimura, On purely transcendental fields of automorphic functions of several complex variables, Osaka J. Math., 1 (1964), 1-14.

[Shv]   O.V. Shvartsman, Chevalley's theorem for complex crystallographic groups generated by reflections in the affine space $\mathbb{C}^2$, Uspehi Mat. Nauk, 34 (1979), 249-250.

[Tkn]   K. Takano, Monodromy group of the system for Appell's F_4, Funkcial. Ekvac., 23 (1980), 97-122.

[T-B]   K. Takano & E. Bannai, A global study of Jordan-Pochhammer differential equations, Funkcial. Ekvac., 19 (1976), 85-99.

[Tu]   Takeuchi, Commensurability classes of arithmetic triangle groups, J. Fac. Sci. Univ. Tokyo, 24 (1977), 201-212.

[Ter 1] T. Terada, Probleme de Riemann et fonctions automorphes provenant des fonctions hypergéometriques de plusieurs variables, J. Math. Kyoto Univ., 13 (1973), 557-578.

[Ter 2] T. Terada, Quelques propriétes géometrique de domaine de F_1 et le groupe de tresses colorées, Publ. RIMS, 17 (1981), 95-111.

[Ter 3] T. Terada, Fonctions hypergéometriques F_1 et fonctions automorphes I, J. Math. Soc. Japan, 35 (1983), 451-475.

[Ter 4] T. Terada, Fonctions hypergéometriques F_1 et fonctions automorphes II, Groupes discretes arithmetiquement définis, J. Math. Soc., Japan 37 (1985) 173-185.

[Thu] W. Thurston, The geometry and topology of three-manifolds, Princeton: Princeton Univ. Press (mimeographed notes 1978-1979).

[T-Y] S. Tokunaga & M. Yoshida, Complex crystallographic groups I, J. Math. Soc. Japan, 34-4 (1982), 581-593.

[W-W] E.T. Whittaker & G.N. Watson, A course of modern analysis, Cambridge Univ. Press (1972).

[Y-Y] T. Yamazaki & M. Yoshida, On Hirzebruch's examples of surfaces with $c_1^2 = 3c_2$, Math. Ann., 266 (1984), 421-431.

[Yau] S.-T. Yau, Calabi's conjecture and some new results in algebraic geometry, Proc. Nat. Acad. Sci. USA 74 (1977) 1798-1799.

[Yos 1] M. Yoshida, Discrete reflection groups in the parabolic subgroup of SU(n,1) and generalized Cartan matrices of Euclidean type, J. Fac. Sci. Tokyo Univ., 30 (1983), 25-52.

[Yos 2] M. Yoshida, A note on orbifold-uniformizing differential equations, Mem. Fac. Sci. Kyushu Univ., 39-2 (1985), 189-195.

[Yos 3] M. Yoshida, Graphs attached to certain complex hyperbolic discrete reflection groups, Topology, 25-2 (1986), 175-187.

[Y-H] M. Yoshida & S. Hattori, Local theory of Fuchsian systems with certain discrete monodromy groups III, Funkcial. Ekvac., 22-1 (1979), 1-49.

[Y-T] M. Yoshida & K.Takano, On a linear system of Pfaffian equations with regular singular points, Funkcial. Ekvac., 19-2 (1976), 175-189.

Sources of Figures

page

55   H.S.M. Coxeter, Regular complex polytopes: Figure 2.4 A,
Cambridge Univ. Press (1974).

56   M. Fujiwara, Theory of ordinary differential equations
(in Japanese): Figures on page 200,
Iwanami Shoten (1930).

57   W. Magnus, Noneuclidean tesselations and their groups:
Figures 17, 18 and 23,
Academic Press (1974).

128   T. Höfer, Ballquotienten als verzweigte Überlegungen
der projektiven Ebene, Figures on page 64,
Dissertation Bonn (1985).

MIX
Papier aus verantwortungsvollen Quellen
Paper from responsible sources
FSC® C105338

If you have any concerns about our products,
you can contact us on
ProductSafety@springernature.com

In case Publisher is established outside the EU,
the EU authorized representative is:
**Springer Nature Customer Service Center GmbH
Europaplatz 3, 69115 Heidelberg, Germany**

Printed by Libri Plureos GmbH
in Hamburg, Germany